MAKING WAVES

Masters of Modern Physics

Advisory Board

Dale Corson, Cornell University
Samuel Devons, Columbia University
Sidney Drell, Stanford Linear Accelerator Center
Herman Feshbach, Massachusetts Institute of Technology
Marvin Goldberger, University of California, Los Angeles
Wolfgang Panofsky, Stanford Linear Accelerator Center
William Press, Harvard University

Series Editor

Robert N. Ubell

Published Volumes

The Road from Los Alamos by Hans A. Bethe
The Charm of Physics by Sheldon L. Glashow
Citizen Scientist by Frank von Hippel
Visit to a Small Universe by Virginia Trimble
Nuclear Reactions: Science and Trans-Science by Alvin M. Weinberg
In the Shadow of the Bomb: Physics and Arms Control
    by Sydney D. Drell
The Eye of Heaven: Ptolemy, Copernicus, and Kepler
    by Owen Gingerich
Particles and Policy by Wolfgang K.H. Panofsky
At Home in the Universe by John A. Wheeler
Cosmic Enigmas by Joseph Silk
Nothing Is Too Wonderful to Be True by Philip Morrison
Arms and the Physicist by Herbert F. York
Confessions of a Technophile by Lewis M. Branscomb
Making Waves by Charles H. Townes

MAKING WAVES

CHARLES H. TOWNES

The American Institute of Physics

Copyright and permissions notices for use of previously published material are provided in the Acknowledgments section at the back of this volume.

AIP Press
American Institute of Physics
500 Sunnyside Boulevard
Woodbury, NY 11797-2999

Library of Congress Cataloging-in-Publication Data

Townes, Charles H.
    Making waves / Charles H. Townes.
        p.   cm.—(Masters of modern physics; v. 14)
        Includes bibliographical references and index.
        ISBN 1-56396-334.5
        1. Townes, Charles H.   2. Inventions.   3. Discoveries in science.
        4. Research—Social aspects.   5. Physicists—United States—Biography.
        I. Title.   II. Series.
        QC16.T65A3   1994   94–28606
        507—dc20   CIP

This book is volume fourteen of the Masters of Modern Physics series.

Contents

SPIRITUAL VIEWS FROM A SCIENTIFIC BASE

About the Series

Masters of Modern Physics introduces the work and thought of some of the most celebrated physicists of our day. These collected essays offer a panoramic tour of the way science works, how it affects our lives, and what it means to those who practice it. Authors report from the horizons of modern research, provide engaging sketches of friends and colleagues, and reflect on the social, economic, and political consequences of the scientific and technical enterprise.

Authors have been selected for their contributions to science and for their keen ability to communicate to the general reader—often with wit, frequently in fine literary style. All have been honored by their peers and most have been prominent in shaping debates in science, technology, and public policy. Some have achieved distinction in social and cultural spheres outside the laboratory.

Many essays are drawn from popular and scientific magazines, newspapers, and journals. Still others—written for the series or drawn from notes for other occasions—appear for the first time. Authors have provided introductions and, where appropriate, annotations. Once selected for inclusion, the essays are carefully edited and updated so that each volume emerges as a finely shaped work.

Masters of Modern Physics is edited by Robert N. Ubell and overseen by an advisory panel of distinguished physicists. Sponsored by the American Institute of Physics, a consortium of major physics societies, the series serves as an authoritative survey of the people and ideas that have shaped twentieth-century science and society.

Preface

I've always been fascinated by the natural universe. Its wonders range from the submicroscopic and most minute—the workings of atoms and the successively smaller particles which make them up—to the grandiose. Why and how is there a universe at all, how did it begin and what are its boundaries? There is also the immediate aesthetic appeal of the beauty of stars, of the sunset, of mountains, deserts, and flowering fields. There is the beauty of remarkable logic in the universe and the surprise that humans can think through and figure out at least some of nature's laws and behavior, many of which, though apparently relatively simple, can predict, and in some sense explain, the remarkable structure and behavior of what we find around us. Most elements of the universe, in fact, seem remarkably simple when understood, yet they synthesize fantastic structures and events. I've always liked solving puzzles. Solving the puzzles of how things around us may actually work, from the completely invisible and yet convincing sub-microscopic structures, to the almost unimaginably large galaxies and universe where they live and move, seems to me the most exciting occupation imaginable.

Fortunately, society has supported this interest in science and I have been able to spend much of my life in the happy occupation of trying to figure out and understand things. Experiments done in the laboratory may deal with situations which are understood well enough that we can predict, check, and see that physical laws hold to the fantastic precision which we can frequently achieve in measurement. Thus, some of my work has involved pressing on toward the highest possible precision, where one can check theoretical prediction of the laws of physics to a part in a million, a part in a billion, and further. But, there are also apparent contradictions. And there are the complex situations such as weather, the formation and distribution of stars, or the events of life itself where some types of predictions can be made and there may be a qualitative under-standing, yet many specifics are unpredictable.

The discoveries of science leading to understanding about how things work and how they can be made to work inevitably lead to many proc-esses and devices which provide possibilities for improvement of human life, or, alternatively, the misuse of such knowledge by humans. While my own interest is primarily in understanding, as is the case for most scien-tists, some of my exploration has led to ideas that are useful to society. Perhaps, most important was the development of a new way to produce radiation represented by the maser and the laser. These have many appli-cations and possibilities, a number of which are now realized and change what humans can do. Some of these now-realized possibilities I could foresee as my work progressed, but many others were initially unpre-dictable even by my natural optimism. For many scientists, research leads naturally to applications of science, and hence also a responsibility to the public in trying to see that modern science and technology are appropri-ately used. Partly for this reason, I have been drawn into public service from time to time and, in some cases, have accepted governmental or ad-ministrative assignments which some scientists have preferred to avoid.

A common public view of the researcher is that of a lonely figure working intensively by himself in a specialized area. But, in fact, science and its applications benefit strongly from the interplay and interaction be-tween scientists, or between scientists and engineers, partly because ideas are clarified and extended by interactions between individuals. Such inter-actions enlarge and accelerate both ideas and the recognition of their ap-plications to the wide variety of fields which are likely to benefit from any really new discovery. I have been fortunate to have enjoyed such interac-tions with the many gifted students and younger physicists who have worked with me, and with stimulating colleagues. Personally I owe much to them.

Although the growth of science depends heavily on stimulating interac-tions within the scientific community, my own tendency has been to avoid staying in popular fields. I like finding the unexpected rather than explor-ing areas which are already widely recognized to be important. I like to think about what beautiful, wonderful, and as yet unappreciated areas may be entered by some new method or exploration. Partly for that rea-son, my own work has moved from one field to another as areas in which I worked have opened up and become popular. At the same time, each move I have made has grown out of what I have experienced in the past, so that there is continuity even when it is not obvious.

The universe is full of logic, which makes even more poignant its mys-teries and puzzles. How its wonders can exist at all, or allow the coherent development of life and of human understanding, is remarkable regardless of how one views them. Its existence naturally challenges our philosophical ideas and comes close to religious thought. To many in the past, science and religion have challenged each other in an unfriendly way, yet to me the challenge is a friendly one, and one that is likely to be resolved in the long run, by unity.

These selections come from different phases of my work and life. They represent things as I saw them and worked through in my career—one with plenty of mistakes, yet some successes, and one in which I have always felt very fortunate.

PHENOMENA OF DISCOVERY AND INVENTION

Quantum Electronics and Surprise in Development of Technology

The evident importance and the considerable expense of scientific research stimulate frequent efforts to assess its contributions to our society, and to optimize its planning. Such efforts are usually undertaken on the premise that we can and should make decisions about the support of scientific research on the basis of what we foresee as its tangible contributions to the nation. While hard-nosed assessment of the contributions of research is clearly appropriate and worthwhile, I am convinced that devotion to this premise is often self-defeating, as will be illustrated here by the obstinate and sometimes bruising facts of past experience.

If we forget the cultural values of knowledge, and evaluate science only by the touchstone of "practical" results, we may at first seem to have a straightforward guide for planning research. We know well that basic research develops many of the new ideas and new information from which technology is derived. Hence, it is easy to conclude that we need primarily to consider what types of technology are wanted for the future, and sponsor those forms of basic science which will contribute the background of information needed for them. There is indeed some truth in this reasoning; it applies particularly to those aspects of technology and of science which we now understand reasonably well, and where we are looking primarily for upgrading of our present abilities or the maturation of developments which are now predictable. But our ability to foresee the practical effects of science is too imperfect. For periods of time as long as a decade or more, or for the really new ideas and startling developments which are not now foreseen, the above approach is unhappily limiting and misleading. Furthermore, it gives a very unrealistic view of the environment needed for high-quality scientific research and of the complex interplay between science and technology, which includes the stimulation of basic science by applied science as well as the reverse.

How, in fact, can we plan for the new idea and the startlingly new, but now unrecognized, technology? Certainly we cannot show that a particular line of basic research will lead to new technological developments if we don't yet even know the nature of these developments. Nor is it possible to satisfy a persistent doubter that present basic research, even though it may be uncovering new knowledge and new ideas, will lead to important though unknown developments for human welfare. Perhaps the best way to examine such questions with some objectivity is the historical method, use of experience.

A general conclusion which seems to me to emerge from a historical approach—the examination of a number of research case histories—is that mankind consistently errs in the direction of lack of foresight and imagination. We continually underestimate the power of science and technology in the long term. Eminently knowledgeable planners and scientists, in attempting responsibly to make realistic appraisals of research, and facing what is at the time uncertain or unknown, all too frequently fall short in foresight and imagination. The element of surprise is a consistent ingredient in technological development, and one we have great difficulty in dealing with on any normal planning basis. Let me now proceed to discuss a particular example with which I happen to be well acquainted—quantum electronics. This is done in some detail, because very specific examples rather than generalities are probably necessary to overcome our natural tendency toward complacency.

Origin of Quantum Electronics

Quantum electronics became a field of physics and of engineering with development of the devices known as the maser and the laser. They use a new type of amplification, the stimulated emission of electromagnetic waves from atoms or molecules. The two devices are of the same general class; in fact, the laser was originally called an optical maser, although the name maser is sometimes restricted to molecular amplification in the radio or microwave range because it was derived as an acronym for *m*icrowave *a*mplification by *s*timulated *e*mission of *r*adiation. The word laser simply means *l*ight *a*mplification by *s*timulated *e*mission of *r*adiation, an application of the same idea to light waves. The parent device involved an amplification technique so radically different that it could not grow out of previous electronics in any orderly way; in fact, its birth in the early 1950's seems to have almost required prior development of the field of basic research known as microwave spectroscopy. How can I justify such a bald statement? Because the idea for maser amplification originated independently in three different laboratories of microwave spectroscopy, and from research rather universally eschewed in applied laboratories. Each of these three origins had a slightly different timing, and differed appreciably in its completeness and practicality. However, all three came from physicists occupied with basic, university-type research on the microwave spectroscopy of gases.

Technology as a Source of Basic Science

It is almost equally significant that microwave spectroscopy itself grew out of wartime technology. This, as well as a good deal of closely related radio-frequency spectroscopy, originated with physicists who had acquired experience in electronics during World War II. In particular, microwave spectroscopy—the study of the interaction between microwaves and gaseous molecules—came about because microwave oscillators and technology were well enough developed during the war to allow this new branch of physics to be fruitful. Thus, a field of basic research was made possible by technology, and the first work in microwave spectroscopy in this country was largely carried out in industrial laboratories. Four independent groups of scientists in the United States, at the Bell Telephone Laboratories, at Westinghouse, at the RCA Laboratories, and at Columbia University initiated more or less independently the study of gases by means of microwaves immediately after the war, and pursued it with some vigor because of its evident importance to physics. The historical importance of technology to its origin is quite clear when one finds that the only university group of these four had been heavily involved in microwave technology during the war and initiated its work to solve an important radar problem. A little later than these four laboratories, the General Electric Company and several universities began further work in the field.

Migration of Microwave Spectroscopy to the Universities

No doubt in the industrial laboratories there was some hope that the new field of physics would have a worthwhile contact with commercial applications. In the case of the Bell Telephone Laboratories, I had myself written a memorandum with some care to convince research management that this could be the case. However, after several years this type of work died out in the four industrial laboratories where it had an early start and moved to the universities entirely. There it attracted a good number of excellent students, as well as experienced professors, because of the insight it afforded into molecular and atomic behavior. Reasons for growth of the field in universities may seem natural enough. Reasons for its decay in industry are equally important, and illustrate rather clearly our dilemma in the planning of research.

Evidently the four large industrial laboratories, although deeply involved with electronics, did not feel at the time that research on the microwave spectroscopy of gases had much importance for their work. I do not know the detailed reasoning of management at Westinghouse and RCA, but after the small teams of research workers which had been quite successful at these laboratories left or lost interest, research in the field was not rebuilt. At the General Electric Company, the research scientist in this field was transferred by management decision to another field considered more pertinent to the company's business. In the case of the Bell Telephone Laboratories, there was a management decision that, while one senior scientist could be appropriately supported, the work was not important enough to the electronics and communications industry to warrant adding a second one. Yet it was out of just this field that 2 or 3 years later a completely new technique of amplification was born which now occupies hundreds of scientists and engineers in the same laboratories. Clearly, misjudgment of its potential was not a simple human fault of any one company or individual; it was a pervasive characteristic of the system.

Sociology of the Maser Invention

Microwave spectroscopy in the universities utilized some of the new electronics techniques of the time, and was able to examine delicately and powerfully the various types of interactions between electromagnetic waves and molecules in ways which were different from those of normal spectroscopy. My own work, by then at Columbia University, flourished in an environment where a considerable amount of related radio frequency spectroscopy was being carried out, and supported by a rather far-sighted Armed Services contract. The resulting development of ideas, in close association with electronics, led in 1951 to invention of the maser at Columbia, and shortly after to other proposals for use of stimulated emission for practical amplification—one at the Lebedev Institute in the Soviet Union and another at the University of Maryland. It is worth noting that basic research in the Soviet Union was at that time primarily concentrated in laboratories of the Soviet Academy, some of whose scientists taught in universities, and that this closest equivalent to our university research laboratories was the setting for the invention there.

By 1954, collaboration with J. Gordon and H. Zeiger produced the first successful oscillator with the new amplifying principle. While a few applied scientists were enthusiastic, overall it evoked only very mild industrial interest. I cannot claim that foresight of the academic community concerning the maser was remarkably greater than that of industrial organizations. But what was important was one of the crucial strengths of academic institutions, that an individual professor by and large makes his own decisions as to what is worthwhile and what might work. This, I believe, generally allows a scientific diversity and utilization of individual insights or enthusiasms in the academic world that are difficult to match in more closely planned and ordered industrial organizations. The latter are especially adapted for a concerted attack on a well-recognized goal. But the diverse and novel ideas for strikingly new approaches to problems are more normally current in communities where vigorous basic research flourishes. Coherent amplification by stimulated emission of radiation, and the idea of gradual quantum transitions rather than quantum jumps, for example, were reasonably well-recognized processes in some academic circles. Applied scientists were at the time characteristically surprised by them. Furthermore, even though there are now many varieties of masers, for some reason the two most complete original suggestions for practical maser systems, from Columbia University and from the Lebedev Institute, both involved molecular beams and Stark effects, techniques and ideas which were of some currency in academic circles but scarcely ever considered in industrial laboratories. But certain ideas of electronic engineering were important too, for example, in providing an understanding of regeneration and of the utilization of coherent amplification. It was the mixture of electronics and molecular spectroscopy inherent in the field of microwave spectroscopy which set appropriate conditions for invention of the maser.

The new type of amplification immediately produced an interesting oscillator, but not so immediately a very usable amplifier. My visit with scientific colleagues at the Ecole Normale Supérieure in Paris generated what seemed to me the first clear view of a practical amplifier by the use of paramagnetic solid materials, because there I was associated with other physicists studying paramagnetic materials and became aware of some of their properties which were otherwise unknown to me. A somewhat similar idea grew up independently from Professor Strandberg, a microwave spectroscopist at M.I.T. He passed on an interest to Professor Bloembergen of Harvard, who had been studying paramagnetic properties for some time, and who provided the variant of the maser which is now its most practical form for amplifiers. By this time industrial laboratories had become more alert to the new possibilities, and it was Feher, Scovil, and Seidel at the Bell Telephone Laboratories who first built a workable amplifier with paramagnetic materials. From this point on, the nation's applied laboratories pursued maser amplifiers for the microwave region with vigor and success.

The Laser

By 1957, I was eager to try to push the new technique on into the shorter wavelength regions, since it was clear that molecules and atoms had the capability of amplifying wavelengths very much shorter than anything previously done by vacuum tubes. I discovered that my friend Arthur Schawlow, then at the Bell Telephone Laboratories, had also been thinking along somewhat similar lines, and so we immediately pooled our thoughts. It was he who initiated our consideration of a Fabry-Perot resonator for selection of modes of the very short electromagnetic waves in the optical region. This very likely had something to do with the fact that Schawlow had first been trained as a spectroscopist and had done his thesis with a Fabry-Perot, another important technique current primarily among university spectroscopists. From this collaboration came the first fully developed ideas for lasers.

The new device was so far out of the normal tradition that its value for applied work was not immediately obvious to everyone. Bell's patent department was at first hesitant to patent our amplifier or oscillator for optical frequencies because, it was explained, optical waves had never been of any importance to communications and hence the invention had little bearing on Bell System interests. But the potentialities were soon sufficiently clear that a number of laboratories in both universities and industry became strongly interested in the optical maser, later called a laser. In particular, not so much later management at the Bell Telephone Laboratories gave it considerable priority. The first actual operating system, the ruby laser, was produced by Maiman at the Hughes Research Lab; this was followed shortly by a second one made by Sorokin and Stevenson at IBM and a third of quite a different type by Javan, Bennett, and Herriott at the Bell Telephone Laboratories. Clearly, the nation's powerful industrial laboratories had begun their push to develop the field.

The successive ideas for improvement and extension of the new type of amplification to the point which I have described came primarily from the realm of basic research. Some of them were rather new, some of them older ideas which had been current in laboratories of basic research. Their sources were almost exclusively scientists trained in microwave and radio frequency spectroscopy. In fact, all but one of those I have mentioned or alluded to above had extensive experience in this field. The demand for such personnel in industrial and governmental laboratories by the early 1960's was, of course, intense.

Practical Uses

What has come out of this development? A total variety of applications too long to list. Since the new technique allows amplification and control of electromagnetic radiation in the infrared, optical, and ultraviolet regions approximately equivalent to what electronics has provided in the radio region, one needs only to think of the utility of light and of electronics to see that a marriage of these two fields would have possible applications in almost any sophisticated technology. I shall give a few examples.

Maser-type amplification comes very close to providing the ideally sensitive amplifier, which can successfully amplify one quantum of radiation. For microwaves, the new amplifier actually provided a sensitivity about one hundred times better than what had previously been available. While by now there are some other types of improved amplifiers, the maser amplifier remains and will likely remain for all time our most sensitive detector of microwaves. Its use is particularly important in allowing efficient transoceanic commercial communications through satellites, scientific measurements of new sensitivity, and in making practical space communications throughout the solar system.

The constancy of atomic properties and the lack of noise fluctuations also make a maser oscillator the world's most precise clock. A maser based on hydrogen is so constant that if kept going for 300,000 years, its expected error would be only about 1 second.

Since light waves can be amplified by the new techniques, they can provide light of almost indefinitely high intensity. Already lasers produce light many millions of times more intense than what was previously available. Laser beams can be accurately controlled and focused to drill holes in refractory materials such as diamond, to partially evaporate and thus precisely adjust electronic circuit elements, or to do delicate surgery. As a surgical tool, the laser is particularly useful in the performance of operations inside the eye without any external incision.

The laser allows our most accurate measurement of distance. In the laboratory, it has detected changes of distance as small as 1/100,000 the diameter of an atom. The coherence of laser light allows interferometric measurements to a precision of a fraction of the wavelength of light up to distances of many miles. This is being used for detection of earthquake phenomena, and for very precise machining. The directivity of laser beams makes them convenient tools for civil engineering; they have been introduced for the boring of tunnels, the dredging of channels, and the grading of roads.

In photography, the new intensities of light available have allowed much higher-speed photography than was previously possible. But still more spectacular is use of the laser as the basis for holography. Laser light projected through a photographic film, with holographic techniques, gives a real three-dimensional image with a wealth of detail and a remarkable depth of focus.

Other uses of laser beams include radar, guidance for the blind, information processing, information storage and retrieval, wireless power transmission, large-screen color television, and cheaper communications.

The Research Planner's Problem and the Drive for Practicality

Consider now the problem of a research planner setting out, prior to invention of the maser, to develop any one of these technological improvements—a more sensitive amplifier, a more accurate clock, new drilling techniques, a new surgical instrument for the eye, more accurate measurement of distance, three-dimensional photography, and so on. Would he have had the wit or courage to initiate for any of these purposes an extensive basic study of the interaction between microwaves and molecules? The answer is clearly No. For a more sensitive amplifier he would have gone to the amplifier experts who, after considerable effort, might have doubled the sensitivity of amplifiers rather than multiplied it by a hundred. For a more accurate clock, he probably would have hired those experienced in the field of timing; for intense light, he would have sought out and supported a completely different set of scientists or engineers who could hardly have hoped to have achieved an increase in intensity by the factor of a million or more given by the laser. For more accurate measurements or for better photography, he would have tried other improvements of known techniques and very likely have achieved moderate success, but no breakthrough by orders of magnitude. It was the drive for new information and understanding, and the atmosphere of basic research which seems clearly to have been needed for the real payoff.

There is at least a superficial similarity between the search for new technology and the pursuit of happiness, each of which is sometimes best approached by indirection. We know some straightforward, but limited, ways to achieve happiness. A better house to live in, or even just an ice cream cone now and then will help. But generally the direct and continuous pursuit of happiness itself is much less successful in achieving the big result than dedication to worthwhile human values and enterprises, without such overt thought of self-satisfaction. Similarly, while direct and planned development of technology is clearly useful and should not be neglected, efforts confined entirely to this approach will be badly limited. Success can be enormously increased by the stimulation and the discoveries which come from an interested dedication to knowledge and discovery themselves.

Americans are intensely practical, and it is difficult to accept the idea that a result is not best achieved by systematic planning, keeping one's eye on the ball, and good hard work. But we have all too frequently had the experience that in judging the practical value of specific scientific research, and in certain cases even of engineering development, those who would seem to be most knowledgeable and responsible are not able to foresee the most imaginative and important steps. History shows this in many more cases than in quantum electronics. In fact, surprise in the development of technology is our regular fare.

Surprise and Nuclear Energy

Some of the interesting story of the development of nuclear energy is quite familiar. Einstein's deduction of the equivalence of mass and energy should have given some inkling of the possibilities even early in this century. During the first part of the 1930's, the exciting field of nuclear physics opened up and produced a small flurry of speculation about the possibility of nuclear energy. But the *New York Herald Tribune* of 1933 carried an assessment of these possibilities under the headline "Lord Rutherford Scoffs at Theory of Harnessing Energy in Laboratories." Rutherford could perhaps fairly be called the greatest experimental physicist of the day and the father of nuclear physics. He had just spoken in Great Britain about the splitting of the atomic nucleus in the same hall where a generation earlier Lord Kelvin, a great physicist of his day, had pronounced the atom indestructible. Rutherford commented, "The energy produced by breaking down of the atom is a very poor kind of thing. Anyone who expects a source of power from the transformation of these atoms is talking moonshine." Professor Rabi of Columbia University, interviewed at the same time, confirmed Rutherford's calculations and hence, apparently, his general conclusions. Professor La Mer, also of Columbia University, was quoted as saying, "I am pleased to see Lord Rutherford call a halt to some of the wild, unbridled speculation in this field." There were indeed some other opinions. Of those interviewed by the *Tribune*, Professors Sheldon of New York University and E. O. Lawrence of the University of California still held out some hope. However, the generation of nuclear energy was not for a few years taken very seriously by the scientific community and was hardly an issue in the support of the study of nuclear physics. In fact, there was considerable concern among physicists, planners, and in industrial circles that too much of physics was swinging toward the nuclear field and that there was too much attention given to this esoteric, relatively useless, aspect of physics. The General Electric Company, deeply involved in power generation, made an overt management decision during this time that the promise of atomic power was not worth its initiating any nuclear research.

Only five years after Rutherford's pronouncement, the unlooked-for phenomenon of fission was discovered, and suddenly the whole world of physics saw the possibilities of nuclear energy in a completely different light. Success could not be assured, but there were now straightforward ways of attempting to obtain nuclear energy. The basic knowledge and knowledgeable personnel were fortunately available because of the previous years of intellectual curiosity centered in the universities; this background and help from Europe's intellectuals were crucial to the United States and its allies.

Other Case Histories

The transistor, another outstanding technological triumph, is by contrast quite a different case, and represents one of success in research planning. M. Kelly of the Bell Telephone Laboratories did foresee that solid-state physics was important in a variety of ways to operations of the Bell System, and formed and encouraged a group of physicists interested in basic exploration of this field. At least initially, this was not done with any direct thought of transistor-type amplification. But it was Kelly's plan of basic physical research on solids, in contact with engineering interests and considerably in advance of most other industrial laboratories of the time, which led to the transistor and its many descendents.

An interesting example of our difficulty with foresight and imagination in a more engineering domain, and where the basic physical phenomena were rather well known, is the case of the airplane. Lord Rayleigh, one of the greatest physicists of the 19th century and certainly familiar with appropriate fields of physics, commented in 1896, "I have not the smallest molecule of faith in aerial navigation other than ballooning." This was followed by severe congressional criticism over the "waste" of government money on Langley's attempts to build a heavier-than-air machine, and was just seven years before the Wright brothers successfully "navigated" over the sands of Kitty Hawk. One can trace an interesting and intense argument for some time thereafter over whether or not the airplane would ever amount to much. Eventually, human need for a flying machine and the characteristically surprising power of technology won handsomely again.

Which Way Genuine Realism?

The above shows us some of the cases where hard realism wasn't real and dreams were. One might well wonder how we can possibly hope to judge the value of specific basic research for the future of technology, and hence on what we can base our plans. My belief is that knowledgeable and responsible people, in trying to judge carefully and not run too much risk of being wrong, have almost inevitably been too shortsighted. Furthermore, planners, in trying to be realistic and faced with tough budgetary decisions, all too frequently find themselves convinced only about what can be demonstrated, and hence their programs are unhappily limited. Science fiction and human need seem to have frequently been more reliable guides to predicting long-range technological developments than sober scientific statesmen. The progress of technology to a point further than we can see clearly—and this means hardly more than a decade—is always surprising and almost invariably greater than we think.

How can we best foster discovery and useful invention? I certainly would not want to play down the importance of planned research and development toward the shorter-term goals which can be foreseen. For this, organized teams and keeping one's eye on the ball can be very effective, and in some cases are almost essential. On the other hand, an atmosphere where utility is paramount is likely to confine thinking in particular channels, and is too prone to smother and draw attention away from what will produce many of the happy technological surprises and radically new ideas. I can suggest three useful guides.

1) There should be an environment of evident devotion to knowledge and discovery themselves, as well as to practical results.

2) To take best advantage of man's curiosity and his potential for discovery, we must give clear attention to supporting the clever, productive, and dedicated researcher in his own insights about what is interesting or fruitful.

3) If the nation is to ensure itself against missing the most exciting surprises, it must ensure support for those fields, even the nonutilitarian ones, where new understanding (not just new detailed knowledge) is most rapidly developing.

I have purposely concentrated attention on the material results of science, but must at least pause to recognize that this involves the frequent mistake of omitting almost completely other important and perfectly real aspects of science and knowledge—their cultural values. Man's view of his universe and of himself which results from scientific research has a significance considerably beyond what is considered "practical" in the narrow sense. Discovery and understanding give breadth of view and inspiration, the satisfaction of man's innate wonder and intellectual drive, and a sense of creative achievement toward some of his most universal goals and most lasting monuments. As something of a parallel to the limitation of being concerned only with the tangible results of science, consider how far short we would be in explaining the importance of music to mankind in a discussion confined to its practical and economic results. However, basic scientific research does, of course, have a profound effect on man's material productivity and wellbeing, and this can be appropriately discussed as long as we remember that there are also other values at stake.

The Short and the Long Run

We have done well in basic research and the generation of new ideas. However, I am genuinely concerned about what seems to me a trend in the United States toward emphasis on the shorter-range goals and over-concentration of attention on utility to an extent which may well limit our technological productivity and leadership in the future. Having emphasized man's limitation in predicting the outcome of research, I do not want at this point to try predictions myself, other than to affirm the continuity of history and the constancy of human nature. However, it is clear that among the many fields where we face decision now are high-energy physics and space exploration. Both are exciting, but expensive. Very little utility can really be predicted for high-energy physics, and little for much of space exploration. Yet we must examine them from both cultural and utilitarian points of view, and with such things on our conscience as the myopic tendencies of the past, our proclivity for taking the lack of foreseeable utility for lack of its real existence, and the ease with which we have disproved the possibility of what only a few years later becomes actuality in this ebullient world of science. And if in these fields or others we are found shortsighted, too lacking in daring, or too indifferent to forward-looking dreams, the pace of science and the impact of technology are now sufficient that our limitations will be obvious not only in the nation's future and the eventual judgment of history, but also to us personally, and in our lifetime.

Origins of the Maser and the Laser

Quantum electronics has its origins in the interplay between different disciplines. I often have thought that the origins of the maser and the laser might well have come much earlier than it did, because I know of no individual theoretical concept that was new in the evolution of those devices. (There were certainly many forerunner investigations that preceded the advent of those devices.) What I think really delayed the development of quantum electronics was a lack of the piecing together of ideas from a variety of fields.

The interaction between technology and science is not a one-way street from basic to applied work, but flows in both directions, from technology to science, and science to technology. The field of quantum electronics might be said to have begun in the "K" band radar development of World War II. I was somewhat involved, in that I was very much concerned about the possibility of absorption of the 1.25-centimeter radiation by water vapor—a concern shared by some others. In looking into that question carefully, I recognized that while it might, and did, deny practical use of radar based on that wavelength, nevertheless it could open up a fascinating field of science, the field of microwave spectroscopy. It was out of this field that quantum electronics really grew. One might suppose that it was just a matter of chance that someone in the field of microwave spectroscopy initiated quantum electronics, but evidently, it was something more than that. There were three apparently independent ideas connected with quantum electronic device origins; one was generated at the Radiation Laboratory of Columbia, the other at the University of Maryland, and the third in the Soviet Union. These ideas were somewhat different in completeness and in timing. However, they apparently were quite independent and all were produced by people working in microwave spectroscopy—three independent starts.

I, myself, initially worked in microwave spectroscopy research at the Bell Telephone Laboratories, which, along with RCA, General Electric, and Westinghouse, had programs in that newly emerging field. (Note that much of that field of investigation was, indeed, initiated in industrial laboratories; that is where the surplus military microwave equipment was.) Now one might think in retrospect, "that is just great, these industrial giants surely recognized that what was going on in microwave spectroscopy might be important to their future." But, in fact, one cannot quite say that about industry. A friend of mine in one of these laboratories was told by his superiors that "the science is fine, but we want you to start measuring the dielectric constants of solids; that is what is important to us." One of the other companies just gradually let the program die, and a third company was sufficiently disinterested to suggest that maybe the gentleman doing that work might like to go to a university, and he did. Bell Laboratories, where I had sufficient support all right, was not, however, interested in extending the field; this, despite the fact that Bell Laboratories is a far-sighted place, and certainly one of the best. But the Columbia Radiation Laboratory already had a start in the field; it had equipment from its World War II research, particularly K-band gear, covering the region of the electromagnetic spectrum that was exciting at that time. When I was invited to come to Columbia with other people who were very much interested in microwave studies, I accepted because of the equipment and the people and the interest there. The Joint Services Electronics Program (JSEP) was backing a liberal program for general and open-minded support of research in the microwave field. I do not think one should criticize industry for dropping research in that field; it is indeed difficult to predict the future payoff of research. What is important is that universities have a special role in carrying out fundamental experimentation, experimentation that sometimes leads to new industry but which sometimes does not. In time, industry played its own role, in that it came very strongly into quantum electronics in the long run. Government has its role too; interaction among these (i.e. universities, government, and industry) really is what we are discussing.

The government played part of its role in the evolution of devices based on quantum electronic principles when I, new to the Columbia Radiation Laboratory, was asked by the Navy to head a committee whose purpose was to examine the possibility of devising new techniques for operating in the millimeter range of the radio spectrum. Moses Long of the Office of Naval Research (ONR) was sort of the activator. I was interested because it was clear that was going to be a rich field for new, high resolution spectroscopy, an extension of my research in microwave spectroscopy. The committee brought together people in a variety of fields that might have some bearing on this effort. Among these were Marvin Chodorow from Stanford, well-known in microwave technology; John Strong, a very well known infrared experimentalist of the time; John Pierce, of Bell Telephone Laboratories, a great expert in electron tubes; and some others. We reviewed the Navy program, trying to develop ideas and suggestions for new things that might generate very high frequency waves. Many interesting ideas were discussed at these meetings. I was working on a variety of things that seemed promising, such as Cerenkov radiation from the surface of dielectric materials, and magnetron harmonics. These techniques worked somewhat, but none of them terribly well. It was not in New York, as someone suggested, but rather in Washington, just before a committee meeting, when the key to getting at the millimeter waves came to me. I was mulling over our lack of real progress, while sitting on a bench in the early morning admiring the azaleas of Franklin Park. I just did not see how one could make the necessary resonators and build the very fine structures with necessary precision, and dissipate the necessary power in them. I had been toying with the possibility of using molecules or atoms as resonators. Clearly, for very high frequency that was the only way to do it, but how? Then I suddenly recognized the possibility. Perhaps the work done at Harvard by Purcell and Ramsey on inversion of populations was in the back of my mind. In any case, I realized there was a way of getting amplification out of molecules and atoms, by avoiding thermodynamic equilibrium. I also remembered a recent talk of the German physicist Paul, who developed a beam technique for obtaining large numbers of molecules or atoms in a single state. On the back of an envelope I quickly worked out a possible way of obtaining an oscillator at high frequency. It seemed difficult, and it was a few months before I could really get anything started on it. Fortunately, there was a good student, Jim Gordon, who was willing to try it. He was joined by Herb Zeiger, a young post-doc who had been working with Rabi and had experience in molecular beam techniques. Although the principle seemed clear, the problem was whether we would be able to get over the margin of oscillation.

The importance of molecular and atomic phenomenon for radio device technology was not an entirely new thought to me. I had, while at Bell Laboratories, written a memo saying that, in the long run, circuit elements at very high frequencies were likely to be using molecular or atomic resonances. But at that time I had not foreseen the possibility of amplification by stimulated emission. Rabi also had pointed out that for very constant frequencies, atomic resonances offered promise. I already had been active in the use of molecular resonances for frequency control, because the Army Signal Corps was interested in those techniques. In that regard, one should note that both Zacharias's cesium atomic clock and the extremely precise hydrogen maser device developed by Prof. Ramsey at Harvard, are products of JSEP sponsorship.

We got our first system oscillating after about three years of work, which is the normal time constant for a graduate student career thesis, and hence the way things tend to move in a university. While we were building it, there was not great excitement about the idea. People said it was a nice idea, but nobody tried to copy it and many were skeptical.

From the beginning, the maser was obviously a good oscillator. I also recognized, at that time, that it also would be a good amplifier; at least it was more sensitive than any previous one, though not very tunable. However, progress took another turn because about that time I took a sabbatical, working in Paris with one of my previous graduate students. In Paris, he and a French physicist had been measuring the relaxation time of certain paramagnetic ions in solids. I found to my surprise that those relaxation times could be very long, and recognized that one hence had the possibility of making an amplifier that was easily tunable and had a much wider bandwidth than the initial maser. We worked on that some in Paris before I moved on to Tokyo, and in the middle of the year made a trip to the Bell Laboratories to talk to the people there about the new device. Note the trail of interaction between different laboratories and people in different disciplines, because that was a very important part of the process. A little later, stimulated by independent work of Strandberg and ours in Paris, Bloembergen had his substantially better idea of a three-level paramagnetic system. By that time I was in Japan, working with one of my previous associates from the Columbia Radiation Laboratory. We wanted very much to understand the noise aspect of the maser. It was easy to show that such amplifiers could get down to the fundamental noise limit of one quantum per unit bandwidth, but the details of the statistics were not clear. Also in Tokyo, I ran into another friend of mine from Columbia who was a biologist. In discussions with him, he brought up a solution to the problem of population changes in microorganisms, which sounded relevant to photon numbers in a maser. I looked up this interesting paper and applied its approach, in somewhat modified form, to noise in a maser amplifier. So, by the time I came back to the United States from Japan I felt we understood the statistics of the noise in the maser amplifiers, including particularly the promising paramagnetic versions.

As soon as I got back to Columbia we devoted our efforts to making a good maser amplifier for radio astronomy. I was interested in radio astronomy; and as Arnold Shostak indicated, the first use of the maser amplifier involved our work with Naval Research Laboratory, for radio astronomy observations using their early radio telescope. It was Kikuchi, a Japanese-American working in Michigan, who first proposed ruby as a good maser substance. This is what we used, and it has been the standard material ever since for maser amplifiers. Subsequently, Arno Penzias, another Columbia student, built another one to look for hydrogen lines. It was later, at Bell Labs, that Penzias and Wilson used a maser in making the important discovery of big-bang radiation.

One might wonder where was the laser all of this time? We were working hard on masers, in the microwave region. In a way, we were distracted because the program was so exciting, and we were enjoying the many new facets of the new high quality oscillators and sensitive amplifiers. Many physicists and engineers were stimulated to build, or suggest new maser devices. It may seem obvious that somebody had to be working on lasers during that time. My own original idea about masers was for getting down into the far infrared; the first one I had written up was to operate at half a millimeter. However, we built the first one in a microwave range, because the microwave techniques were available and easier. I thought to myself that of course we would get down to the submillimeter range or in the far infrared sometime, but it would be a bit more difficult. Sometime during this period—I suppose it was probably late 1956 or early 1957—Bill Otting, of the Air Force Office of Scientific Research, dropped in to see me and asked whether I would be willing to write a review paper or discussion about the possibility of getting into the infrared. That, I think, showed a great deal of prescience on his part. It was typical, at the time, of our relations with military scientists. They would come by and want us to do this or that. Sometimes we did it, sometimes we did not, which they understood. In this particular case, I told Otting I thought it was very interesting and I wanted to do it, but I really did not have any great new ideas over the original one and hence did not feel like there was anything much to write at that point. I did say that when I had additional good ideas I would consider writing such a paper. I also mentioned that there was a very bright young man who had just gotten his Ph.D., namely, Ali Javan, who might be willing to do a write-up on the possibilities. Well, Ali was busy with other things too. He was not inclined to do so at the moment. I believe the first time that anything was written about really getting into the short infrared came from an Air Force study in the summer of 1957. That study was on the future of the Air Force, and what they should be working on. In that report one will find mention of the possibility of pushing masers perhaps as far as the mid-infrared. I was part of that study, and of course, that was the reason it was in there. The report was never issued because it was written in the late summer of 1957. Sputnik came out in October, and the report was outdated by that event. The Air Force had carefully avoided talking about space work in the report because that was unpopular in Congress until after Sputnik, so that report never came out. However, it does exist in the files of the Air Force. In September of 1957 I decided that I really ought to sit down and think about how to get to shorter wavelengths—meaning pushing on down well below the millimeter wavelengths. A little thought convinced me that the gain was just as easy to get in the optical region as in the far-infrared domain. Equations showed that you typically did not need any more atoms in the optical than in the microwave region, and that the gain was just as easy to get in the optical region. Since we knew a lot more about the optical region and optical techniques, I decided we should just jump directly into the optical region rather than initiate work in the far infrared. But one of the real problems in doing that was to design a single-mode resonator. I designed a resonator with some big holes in it that suppressed some of the modes. I was going to bounce some of the exciting radiation back and forth in it and get rid of some undesired modes, but I recognized that there were still many modes present and that radiation would probably hop from one mode to another, Nonetheless, it would be an interesting development to obtain an optical or infrared oscillator. Since at the time I was at Columbia University, one might wonder then, why wasn't the laser a direct JSEP Columbia University invention? The reason was somewhat accidental. I had done the above described work at my desk. But I was consulting once every two weeks, for a day, at the Bell Telephone Laboratories. I went out to the Bell Telephone Laboratories and talked with Dr. Arthur Schawlow, who had done his post doctoral work with me at Columbia, and had then gone to Bell Labs. He said he was very much interested in the idea and had been thinking about such things. As I told him about what I had done, he asked me about a Fabry-Perot resonator, since that would eliminate a lot of the modes. Immediately I recognized that was the real idea for a resonator. We worked out ways in which one could eliminate all but one mode in a Fabry-Perot system. I then had the problem of what was the ethical thing to do—how does one divide the pie here? Was I working for Columbia and JSEP, or was I working for Bell Labs when I was sitting in my office? I recognized that if both institutions were involved it was going to get complicated. Well, I reasoned, Art Schawlow is there at Bell Labs, and he has made a big contribution to the subject; so at least half of it should be a Bell Laboratories invention. So making it a Bell Laboratories patent was a natural way out. However, I think from the point of view of ideas, JSEP has equal claim. It was a completely arbitrary decision on my part to say, "Well, let's just call this a Bell Laboratory device—that is the simplest thing to do." But JSEP was a crucial contributor to the development, including Art Schawlow's and my association. I do not think one should make a great deal about the details of what happened at which institution, because the interaction between institutions was a very important part of the process, and the ease of naturalness of that interaction is something we must try to preserve.

Sometime in 1958, we started working on trying to build a laser at Columbia. A new graduate student, Isaac Abella, was going to undertake this as a project. He was working with us on that project as a thesis problem. However, I got persuaded that it was more important for me to be down in Washington for a couple of years, so in the fall of 1959 I went down to the capital. Isaac was still working at it at that time and surely would have made a laser eventually. But now was the time for industry to shine. Industry already had become excited about masers as amplifiers, particularly the three-level, solid-state maser. There also had been industrial interest in the ammonia maser as a constant-frequency oscillator. Students in microwave and radio astronomy had been hired by industrial groups to work in that field. In fact, the maser business became so exciting that the editor of *The Physical Review* stated that too many people were submitting papers on those devices and he felt forced to put a limit on publication on that topic. But what really ticked things off as far as the lasers were concerned, was the paper by Schawlow and myself. Everybody was ready by then to recognize their excitement and potentiality. Industry got going, and all of the first types of lasers were first made in industry. Almost all of the important actors in quantum electronics came out of the field of radio and microwave spectroscopy developed in the universities. The very first laser was that of Ted Maiman. (Maiman was grandson of the Columbia Radiation Laboratory program, because he was student of Willis Lamb's who had been a major figure there.) Quantum electronic techniques are really applied spectroscopy. Maiman was working at Hughes Research Laboratories, having been brought there by Harold Lyons. Lyons had been involved in the atomic clock business at the National Bureau of Standards and was closely connected both with Zacharias's work and with our ammonia work. The next laser that was built is not so widely known simply because it has not been a commercial product. It was made at IBM by Peter Sorokin and Mirek Stevenson. Sorokin was one of Bloembergen's students, from the Harvard program. Stevenson was one of my students from the Columbia program. Both had been hired by IBM. Together they made the second and third lasers. Those were crystal lasers, interesting lasers at the time but they have not been commercially useful. The next laser was made by Javan, Bennett, and Herriot. Javan had gone from Columbia to the Bell Laboratories. This important device was the helium-neon laser—the most popular laser used today. (It has sold more than any other laser and has been an exceedingly useful laboratory, as well as industrial tool.) Both Javan and Bennett had come from our Columbia program. Herriott was an optical physicist who had been at Bell Laboratories for some time. Javan provided the initial idea and figured out the atomic physics involved in collaboration with Bennett, while Herriott had concentrated more on the optics. Together, the three of them got the first gas system going.

Today, one may note that lasers can be made out of almost anything. Almost anything can be made to amplify, if properly excited. I must say in the early days that would have been surprising. I thought it remarkable that, although spectroscopy had been going on for generations with very highly qualified people working in the field, and a wide variety of experiments had been carried on with gas discharges, that no one had ever seen any kind of amplification. Therefore, I reasoned, the necessary conditions must be pretty tricky. By now, many different varieties have been generated by clever people and they do not seem so difficult.

Even after successful lasers were made, it still took the field a while to get going, in an industrial sense. There was a famous quip around which people used to twit me, that the laser was "a solution looking for a problem." Nevertheless, industry was steadily very active in the field. It seemed an exciting one, and by the late 1960s it began to really see use and find very significant applications. By now there are, of course, an enormous number of important laser uses. Twenty five years after its invention, the laser involved about a 4.5 billion dollar industry. After all, such a device that combines the fields of optics and electronics must eventually touch a wide variety of scientific and technological areas. The many uses of the laser are now well known. This is an almost textbook example of basic research and its contribution to technology, industry, and the military. One might also add the interaction back on the basic research of the now many industrial and technological developments in this field. Few lasers are built by universities these days; the commercial devices do much better than we can do at the universities.

Escaping Stumbling Blocks in Quantum Electronics

I t is sometimes said that there is no single component idea involved in the construction of masers or lasers which had not been known for at least 20 years before the advent of these devices. Of course, a discovery which might have occurred earlier is not uncommon in science or engineering. Nevertheless, the case of quantum electronics is striking enough that it may be useful to review the development of ideas prior to the time this new field became visible, and the stumbling blocks which may have delayed its creation. I believe whatever unnecessary delay occurred was in part because quantum electronics lies between two fields, physics and electrical engineering. In spite of the closeness of these two fields, the necessary quantum mechanical ideas were generally not known or appreciated by electrical engineers, while physicists who understood well the needed aspects of quantum mechanics were often not adequately acquainted with pertinent ideas of electrical engineering. Furthermore, physicists were somewhat diverted by an emphasis in the world of physics on the photon properties of light rather than its coherent aspects. It is still surprising that the basic combinations of ideas required for quantum electronics were not more completely envisaged somewhat earlier than they were. Nevertheless, it is understandable that the real growth of this field came shortly after the burst of activity in radio and microwave spectroscopy immediately after World War II since this brought many physicists into the borderland area between quantum mechanics and electrical engineering.

I know that most of my electrical engineering friends, while well acquainted with absorption of radiation by atoms and molecules, were surprised to learn that excited molecules could give up energy coherently to an electromagnetic wave. With that knowledge in mind, they might at least have imagined utilizing such effects for amplification even if they were not expert with the specific arrangements required. Many physicists knew of stimulated emission, but few connected it with useful amplification. That amplification by stimulated emission had been well understood by many individuals is impressively demonstrated by a number of early records—from Richard Tolman's publication in 1924 discussing amplification by an inversion of population, to serious consideration of an experiment to demonstrate stimulated radio emission by John Trischka at Columbia University several years before invention of the maser. In all, I know of eight apparently independent discussions prior to the maser invention of how stimulated emission and nonequilibrium populations can increase the intensity of a wave. There may be others.

Why were the many discussions of stimulated emission not followed up to produce some actual demonstration or a useful device? In some cases such effects appeared to be only of theoretical interest, providing a neat and consistent explanation for characteristics of radiation and of absorption. Furthermore, the simple weakening of absorption with increasing population in an upper state (e.g., for $kT>h\nu$) always seemed to me an adequate demonstration of stimulated emission. The few experiments on stimulated emission attempted in the optical region made demonstration of any large effects seem difficult, and were not followed up by other experimenters. Probably the failure to couple the idea of feedback with weak stimulated emission helped make such effects seem inevitably small. In the case of Trischka, and in my own thinking before a feedback oscillator was envisaged, a demonstration of amplification in the microwave region due to inversion seemed rather difficult and not important enough to be worth the required time. In the case of inversion of nuclear spins, which was obtained some years before maser action, the low frequencies and nuclear magnetic dipole moments involved made most stimulated emission effects quite small, and presumably for that reason successful inversion of populations did not direct attention toward useful amplification.

In my own case, it was primarily a strong desire to obtain oscillators at shorter wavelengths than those otherwise available that induced me to initiate experimental work on the maser. This is why the first system designed on paper was for 1/2 mm wavelengths, in the far infrared. My students and I had previously tried many techniques—magnetron harmonics, coherent Cerenkov radiation, and others to obtain short wavelengths, and while most of them worked after a fashion, none gave the promise which masers did for good spectroscopic sources at short wavelength. Initially, I did not fully realize the maser's potential as a low noise amplifier or as a precision clock, though these two applications added considerable interest rather soon after work on a maser was started with James Gordon and Herbert Zeiger.

As indicated above, my belief is that for many of the physicists who understood stimulated emission, isolating such effects seemed somewhat difficult, and the necessary experiments were not very important because to these same physicists stimulated emission was already rather well understood. The idea of feedback and large numbers of quanta in single modes which might have suggested practical applications and given additional value to such experiments had not occurred. In addition, there were some misunderstandings and confusions which played a role in delaying quantum electronics. About 1945 I had myself written an internal memorandum at the Bell Telephone Labs explaining that molecules and atoms could be used to generate short microwaves, but that intensities could only be low because they would be limited by the second law of thermodynamics—use of a nonequilibrium distribution and population inversion had not yet occurred to me.

Emphasis on the photon aspect of light deflected some physicists from coherent amplification. It turned out that before the maser was operational, John Von Neumann had suggested exciting electrons in a solid by neutron bombardment and thereby obtaining a powerful cascade of photon emission. Coherence was not mentioned, and I believe no one ever attempted such an experiment. J.H.D. Jensen told me that in the 1930's he had thought about stimulated emission from an inverted population as a cascade of independent photons, like a cosmic ray shower. He lost any great interest in the idea after an experiment which seemed to produce such effects turned out to be explained otherwise. In thinking about light itself, rather than microwaves, it may be that many electrical engineers would have not been any more concerned with coherence effects than were these physicists. However, both engineers and physicists were naturally led to consider coherence when dealing with the radio or microwave region, and I believe this is why initial ideas and development of the field were so dependent on those with experience in radio and microwave spectroscopy.

Some of the confusion about coherence seemed a little strange even in the early days, and will seem remarkable at this time, but was real. Consider the early experiment of A.T. Forrester. Shortly after World War II, he began an experiment to irradiate a photoelectric surface with two Zeeman components of an optical line in order to mix the two frequencies. The idea was to have the two frequency components separated just enough to produce a varying photoelectric current at a microwave frequency low enough to detect by available electronics. Forrester's interest, I understood, was to demonstrate the effect of mixing two discrete optical lines, an effect which should have been detectable by the recently developed high-frequency electronics. The experiment was not easy at the time, and apparently a number of physicists believed it to be conceptually wrong. There seemed to be confusion over the spatial extent of the possible coherence, and questions whether the independent particles of different frequencies could cooperate in emitting electrons from a surface and thus give a beat. My belief then and now is that a bright electrical engineer would have figured out that the experiment would work. Nevertheless, the basic idea was challenged by a publication in the *The Physical Review* in 1948 and by enough scientists that after the experiment was under way I was asked by the sponsoring agency to review it and advise whether the idea was faulty. The experiment was only difficult, not erroneously planned, and Forrester's published result later dispelled any doubts.

Perhaps a somewhat more subtle example of physicists' bent at the time toward thinking in terms of individual particles involved the frequency spread of a maser oscillator. From the point of view of stimulated emission produced by an oscillating field established in a resonant cavity, it is not hard to understand that the radiation produced by a maser oscillator could indeed have a very narrow frequency band, independent of the width of response of individual excited molecules. Any real width has to be due to either the small amount of additional spontaneous emission or to thermal radiation present, as first worked out by James Gordon. However, there was the uncertainty principle relating time and energy, a basic law for physicists. With lifetime t of molecules in the cavity limited (for the beam-type maser) by the time of transit, how could there be a frequency width much smaller than $1/t$? An electrical engineer accustomed to the almost monochromatic oscillation produced by an electron tube with positive feedback would perhaps not have given the problem a second thought. However, before oscillation was achieved I never succeeded in convincing two of my Columbia University colleagues, even after long discussion, that the frequency width could be very narrow. One insisted on betting me a bottle of Scotch that it would not. After successful oscillation, I remember interesting discussions on this point with Niels Bohr and with Von Neumann. Each immediately questioned how such a narrow frequency could be allowed by the uncertainty principle. I was never sure that Bohr's immediate acceptance of my explanation based on a collection of molecules rather than a single one was because he was convinced, or was due simply to his kindness to a young scientist. My discussion with Von Neumann had a more special twist and occurred at a social occasion. When I told him about our maser oscillator, he doubted that the uncertainty principle allowed our observation of such a narrow frequency width to be real. After disappearing in the crowd for about 15 minutes, he came back to tell me he now understood the situation; my argument was correct. That Von Neumann took even that long to understand impressed me. He then went on to urge that I try for stimulated emission effects in the infrared region by exciting electrons in semiconductors. I was puzzled by his strong insistence on this, because at that time the use of semiconductors for stimulated emission amplifiers seemed much more difficult than other possible methods. It was only much later I learned that he had already independently proposed such a system to produce a powerful avalanche of photons by stimulated emission, and must have been avoiding the more usual response of saying he had invented the idea before he heard of our work.

I must note that, although a number of good physicists did not find the coherent aspects of masers straightforward, for most of those in the field of radio and microwave spectroscopy, they were fairly obvious. That was true, for example, of my colleagues I.I. Rabi, Polycarp Kusch, and Willis Lamb, and of course Arthur Schawlow with whom I later collaborated on the laser. Electrical engineers, while not so knowledgeable about quantum properties, also found coherence properties very natural and seemed to have the right instincts about many aspects of quantum electronics from parallels in ordinary amplifiers and circuits. As an illustration, I give an example of how an electrical engineer helped me at one point.

The frequency stability of a maser oscillator was an important question, and I had worked out an expression for it which showed, I thought, that the cavity pulling would give an error proportional to the square of the ratios of quality factor Q for the spectral line to that of the maser cavity. On explaining this to an electrical engineer at Stanford (whose name unfortunately I cannot recall), he remarked that such a result was peculiar since frequency pulling of one resonance by another in circuits was proportional to the first power of this ratio. While denying any detailed knowledge of the quantum mechanical properties, he doubted my result was correct and he was right. A little further examination when I returned home showed that I had neglected reactive terms of the quantum mechanical oscillator and cavity, which then gave just the result he expected.

As of today, the more engineering ideas of coherence, feedback, and nonlinear frequency mixing have become so intermingled with the more physical ideas of discrete states and quantum mechanical processes in the minds of both electrical engineers and physical scientists that some of the above confusions will probably be hard to believe. That is why the few specific examples above may help remind us how things were.

In addition to conceptual stumbling blocks which affected the course of quantum electronics, in the early days there was also a limited apprecia tion of the potential of this field, and this too may deserve illustration. Of course, I do not pretend to have foreseen the field's full potential myself, though I was obviously more impressed by it than many others.

Consistent with my usual practice of working with graduate students, development of the maser proceeded at the rate of a normal graduate student thesis project, being completed by Jim Gordon approximately three years after the idea was conceived. Our laboratories at Columbia University were completely open in the usual way of academic institutions, and many knowledgeable people visited the maser experiment. However, no one seems to have thought it interesting enough to reproduce or to try to compete with us. As far as I know, there were no concurrent efforts to obtain a molecular or atomic oscillator except the work of Basov and Prokhorov in the Soviet Union, and in this case I am not familiar with just what was done in these earliest years.

Completion of the maser oscillator was exciting to some, but evoked no more than mild interest on the part of other of my friends and did not immediately generate any great flurry of work. For some time it appears that the potential of quantum electronics was unappreciated by many of those not already in the field. In part this result could have been because the first maser itself may have been judged both limited and specialized. But also, for understandable reasons, scientists busy with their own research are not necessarily quick to see the potential of new events in other fields. Whatever the reason, when the performance of masers as frequency standards and then as amplifiers became more evident and as new varieties such as many solid-state systems were proposed, interest in masers grew. By 1960, publications on new varieties of masers became so common that, presumably under the assumption that the excitement must be over, the Letters section of *The Physical Review* made public a policy not to accept any more letters on new masers.

The delay of about six years between masers in the microwave region and lasers, or masers at shorter wavelengths, was no doubt also due in part to some conceptual stumbling blocks. One of these was imperfect recognition of the possibility of obtaining a high Q and of emphasizing a single mode in a structure which is very large compared to a wavelength, like the Fabry-Perot, even though Fabry-Perot resonators were well known. This problem and some others made it difficult to recognize ways that masers operating at infrared or optical wavelengths could perhaps be as easy or easier, rather than harder, than those in the microwave region. I shall not here try to explore the missing conceptual links, but rather turn to a different aspect, an apparent lack of appreciation of the potential of optical masers prior to late 1957. A number of individuals certainly recognized that maser techniques might be extended to much shorter wavelengths. I believe it was in 1956 that Bill Otting, Head of Physics for the Air Force Office of Scientific Research, asked me if his office could support me or someone else I might suggest in work towards an infrared maser oscillator. It is difficult to remember how many other more casual conversations there might have been on the subject or to know how many scientists may have considered this, but there were apparently no substantial efforts to explore maser oscillators at wavelengths much shorter than the microwave region before 1957. I know why I myself delayed this long—I was busy with and excited by microwave applications of the maser and saw only rather brute-force methods of moving to much shorter waves before that time. I was waiting for a "neater" idea to occur. About others I have no direct evidence, but believe a lack of appreciation of the potential of lasers and a closely connected effect due to the state of development of optics and optical oscillators both played a role in the time-delay between microwave and optical oscillators.

Optical spectroscopy had its heyday for physicists in the 1920s and 1930s; by 1940 most physicists considered it a mature field of solid importance but from which no remarkable breakthroughs were likely to emerge. There was, I believe, an attitude of déjà vu about optics. After World War II optics and optical spectroscopy did have a substantial renaissance, especially in the hands of French physicists, but was still not an area to which many turned for forefront physics. As I see it, lasers might well have been invented during this 1925–1940 period, although they would have been more difficult for lack of certain techniques which were further developed in later decades. These include a miscellany of things such as good optical coatings, flash tubes, and improved varieties of infrared materials and detectors.

As the possibility of high quality, single or almost single-mode lasers came into everyone's view, interest and intensity of attention to this field increased sharply. Nevertheless, much of its now quite evident potential was initially appreciated only by enthusiasts and some not even by them. There was almost immediate interest in the optical maser proposals of Schawlow and myself, but the beauty of the device may have been more attractive to most scientists than its potential applications. A favorite quip which many will remember was "the laser is a solution looking for a problem." While an enthusiast myself, and aware of the potential for high precision measurements, monochromacity, directivity, and the high concentration of energy that optical masers would provide, I missed many potent aspects. The area of medical applications is one that did not occur to me initially as promising. In retrospect, I can imagine recognizing the beauty of operating directly through the pupil without other insults to the eye, but since I had never heard of a detached retina such an idea would have been another "solution looking for a problem." My own scientific interests were primarily in the direction of new forms of spectroscopy and precision measurement, and hence I needed only modest power. While it was evident that optical masers could be expected to produce powers of at least a number of watts, I did not initially think of very short pulsed operation at a power level of many kilowatts, as was produced by Maiman's ruby laser.

In looking back over why the field of quantum electronics took as long as it did in getting started and why even then the buildup was initially not more rapid, I necessarily mention some of the stumbling blocks, misconceptions, and fumbles. The development of any science by humans has its similar mistakes and illogicalities. Recalling these can keep us humble and make us aware there may be other exciting events not yet visible around the corner. However, focusing on problems of the past omits or deemphasizes the remarkable insights and inventions made by a large number of colleagues who have contributed to this field, and the vigor with which industry pursued and developed it. I can resist discussing these impressive aspects of the field only because I know others will treat them appropriately.

QUANTUM
LEAPS

Microwave Spectroscopy

Microwaves are a variety of electromagnetic radiation with wavelengths ranging roughly between one meter and one millimeter. They are almost the last type of electromagnetic radiation to become well known and useful to human life. A brief review of the other varieties of electromagnetic waves will illustrate their familiarity and importance to man. The 60-cycles-per-second electrical waves which are sent along power lines into our houses exemplify the low-frequency types and have the enormous wavelength of about 3000 miles. The somewhat higher-frequency radio waves used for ordinary radio broadcasts have wavelengths as short as a few hundred feet; the shorter radio waves used for television broadcasts are only a few feet long and border on the still shorter microwaves. Skipping the microwave region for the moment, we come to infrared radiation, which begins with wavelengths of about one millimeter or frequencies of 3×10^{11} cycles per second and extends to wavelengths as short as 0.001 millimeter. Infrared radiation, of course, is important in transmitting heat. Immediately beyond the infrared region is ordinary light—a rather well-known variety of electromagnetic radiation. At still shorter wavelengths there are the ultraviolet and then the X-ray radiations. Beyond the X-ray region are the gamma rays and cosmic rays of interest to the nuclear physicist.

All of these types of electromagnetic waves have been, at least until quite recently, much better known and much more widely used than microwaves. One might have expected that, since microwaves have wavelengths comparable with laboratory dimensions, they would have been the most convenient for study and among the first types of electromagnetic waves to be investigated. In fact they were, for the nature of electromagnetic waves was first-proved by Hertz with the production of microwaves in an oscillating spark discharge. Their study has been long delayed, how-

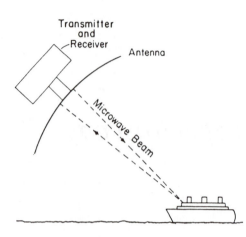

Figure 1. Schematic plan of general operation of a radar system.

ever, primarily because of the difficulty of generating microwaves in a controlled and usable fashion. Longer-wavelength radio waves are easily produced by electronic tubes, shorter-wavelength electromagnetic rays are easily produced by incandescent lamps or other hot objects, but in the microwave region nature has not provided convenient means of generating electromagnetic waves. Although these waves have been produced in a limited way for many years, it was only during the recent war that microwave generators, apparatus, and techniques were sufficiently developed to make their use and study convenient.

Radars and One That Failed

Much of the stimulus for the wartime development of microwave technology came from radar. Figure 1 shows in broad schematic plan how a typical radar system works. The microwaves are generated and emitted by an antenna which is essentially a large parabolic mirror. As in a searchlight, the parabolic mirror sends out a well-defined beam of microwaves. If in the path of this beam there is some object which reflects microwaves, such as the ship shown in the figure, the microwaves are reflected and sent back to the same parabolic antenna which focusses them into a radio receiver, a device for detecting the microwaves. If microwaves are detected when the searchlight antenna is pointed in a particular direction, the radar operator knows that some object of interest is present. The whole device is much like an optical searchlight with, however, several distinct advantages. The first, an advantage only in military operations, is that someone being illuminated by the microwave beam may not know that he is in the path of the beam. A strong optical searchlight would, of course, immediately warn anyone on the vessel that it is being observed. A second advantage, which applies also in peacetime, is that these microwaves can penetrate haze and rainclouds which would completely stop optical light.*

Waves can bend around obstacles only if the obstacles are smaller than the wavelength. Thus, ordinary radio waves easily bend around buildings and mountains, the shorter television waves tend to require an unobstructed view for good reception, while optical waves are so short that they normally travel only in straight lines. The microwaves are long enough to pass around raindrops or fog particles and thus can penetrate a cloud, whereas the short optical rays cannot.

When the United States entered the last World War, the best radar available used microwaves approximately 10 centimeters long, a type developed by British scientists. The 10-centimeter microwaves were extremely useful, but because of their long wavelength could not be focussed into a very narrow beam by the parabolic reflector. The diffraction pattern of such a reflector produces a beam several degrees wide, so that the positions of the objects being observed could not be determined very exactly. After much engineering and scientific effort, the United States succeeded in producing good radar sets with microwaves as short as 3 centimeters. The result was an enormous improvement in accuracy and in the detail which could be seen. The shorter wavelength was so successful, in fact, that intensive work was initiated to produce still shorter wavelengths.

The wavelength for the next improved radar was somewhat arbitrarily chosen as 1.25 centimeters. After about two years of effort on the part of engineers and physicists, as well as an expenditure of millions of dollars, such a radar system was produced, installed in an aeroplane, and tested. The new radar produced a narrow searchlight beam as expected, but the range of visibility was disappointingly small. It could not reach much farther than fifteen miles, whereas 3 centimeter radar had been reaching seventy-five or a hundred miles.

A flurry of research to understand this unexpected trouble soon showed that the 1.25-centimeter microwaves were being strongly absorbed by water vapor in the atmosphere. The results are shown in Figure 2. There is a broad region of absorption due to water vapor for wavelengths between 1.0 and 1.6 centimeters, with the maximum absorption unfortunately very close to the wavelength chosen for the radar—1.25 centimeters.

---

* Another advantage is radar's ability to measure distances accurately by emitting pulses of microwaves and timing the return of reflected waves.

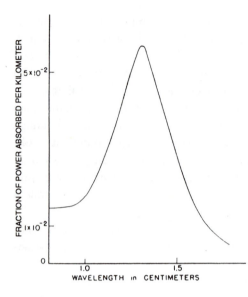

Figure 2. Absorption of micro-waves by water vapor in air. 10 grams of H_2O per cubic meter of air are assumed, or 50% humidity at 23°C.

At this stage of the war the United States was primarily engaged in warfare in the Pacific area, where water vapor in the atmosphere is high; since this water could not very well be eliminated, most of the radars had to be discarded. The parts of these sets were put on shelves or junk-heaps, and we went back to the use of the 3-centimeter radar, which fortunately was still a very successful device. This accidental wrong choice of wavelength was at the time a major military setback but it was a tremendous boon to postwar research. The discarded 1.25-centimeter radar components have been avidly sought out by physicists and chemists for microwave spectroscopy and are the envy of scientists in countries less blessed with discarded radars.

Let us turn now to an examination of the nature of this absorption or response of water vapor to microwaves. All electromagnetic waves are generated or absorbed by moving or oscillating electric charges. Currents move up and down the antennas of a broadcasting station and thus create radio waves. Radio waves are received similarly on antennas in which currents are induced to move back and forth by their action. Light waves are normally produced by electrons rotating or moving very rapidly in atoms. Similarly microwaves can be produced by the rotation of a molecule if this molecule has positive and negative charges associated with it, and it is such a rotation of the water molecule that makes it respond to microwaves.

Consider as a simple example a molecule of table salt, composed of an atom of sodium attached to an atom of chlorine. Three possible types of action can occur in this molecule. Electrons move very rapidly around the

Part of the Electromagnetic Spectrum		
Microwave Spectra	**Infra Red**	**Optical & Ultraviolet**
Wavelengths 10^{2} cm.– 10^{-1} cm.	10^{-1} cm.– 10^{-4} cm.	10^{-4} cm.– 10^{-6} cm.
Sources of Typical Spectra: Molecular Rotation	Molecular Vibration	Electron Motions

Figure 3. Relations among various types of spectroscopy.

chlorine or sodium atom, and their frequencies correspond to the frequencies of light waves. The molecules can also stretch or contract by changing the distance between atoms, so that the atoms vibrate against each other. The entire atoms vibrate more slowly than the electrons because they are considerably heavier, and their vibrations correspond to the longer-wavelength infrared rays. A still slower motion is the end-over-end rotation of the sodium chloride molecule. Since the sodium tends to be positively charged and the chlorine negatively charged, this end-over-end rotation corresponds to a periodically oscillating motion of electricity which can radiate waves of the same frequency. The molecule can also absorb waves which resonate with or have the same frequency as its rotation.

The general relationship between the various molecular motions and various types of spectroscopy is shown in Figure 3. The boundaries between the different spectral regions are of course not sharply defined. There are some vibrational frequencies which are so slow that they lie in the microwave region and some rotational motions which are so rapid that they lie in the infrared rather than the microwave region. In general, however, the rotational motions are characteristic of the microwave region.[*]

It may be noted that each major type of spectroscopy has developed historically out of some particular technique: radiation in the optical region is observed visually; infrared radiation was explored by heat detectors. Microwave spectroscopy is characterized by radio and electronic techniques, and it is the limitations of these techniques, rather than any real change in the nature of radiation, which fix the limits of microwave spectroscopy at one millimeter and one meter. The shortest essentially monochromatic radiation so far produced and detected electronically had a wavelength of 1.1 millimeters; hence the lower wavelength limit of microwave spectroscopy. Usual microwave spectroscopic techniques cannot go beyond one-meter wavelength because the intensity of absorption decreases very rapidly with increasing wavelength and at such long wavelengths is entirely too small to be detected. It has already been noted that the 1.25-centimeter radar wave did penetrate as far as fifteen miles, so

---

[*] Discussion of other interesting types of microwave spectroscopy in solids, liquids, and ionized gases must unfortunately be omitted for the sake of brevity.

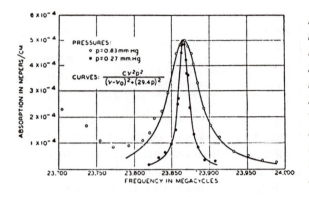

Figure 4. Variation in a microwave absorption line of NH₃ with pressure. The line becomes narrower but the intensity at peak is constant with decreasing pressure. The rise of absorption to the left is due to the presence of another absorption line.

that in ordinary laboratory distances, a small fraction of their power is absorbed—in fact a fraction 10^{-5} per centimeter path. At wavelengths of one meter, absorption by molecules can be expected to be still less by 10^5, or about 10^{-10} per centimeter, and this is too small to be detected by present techniques.

Spectroscopy at still longer wavelengths exists and has been given the names of molecular beam spectroscopy, nuclear resonance, and others. These forms have much in common with microwave spectroscopy and are almost as new. We shall hence be comparing microwave spectroscopy primarily with the better-known spectroscopy in the infrared and optical regions.

Accurately Tuning-in on Molecules

Detailed and accurate determination of the properties of water can hardly be expected to come from the broad band of absorption shown in Figure 2, since no distinguishing feature can be accurately measured. There is fortunately a very easy way of making this response sharper and more accurately defined. The absorption shown in Figure 2 applies to water vapor in air at atmospheric pressure. If the pressure is decreased to one-half of an atmosphere, the width of the absorption band decreases by half. The intensity, however, stays essentially constant. If the pressure is decreased further, the width continues to decrease but the amount of absorption at the maximum stays constant, as shown in Figure 4. This is remarkable behavior and contrary to experience in any of the other spectroscopic regions. It holds true for a decrease in pressure of about one million. The absorption band—now so narrow that it might be called an absorption line—decreases, as a result, by a factor of one million without any appre-

Figure 5. Pure rotation spectrum of a simple molecule.

ciable decrease in the amount of absorption at the center of the line. The range of absorption becomes at one-millionth atmosphere so narrow that it could only be represented on Figure 4 by zero width. Under these conditions the frequency of the molecular response can be measured with an error less than one part in 10^7.

With such a sharp response, the molecule might be considered a very highly tuned radio receiver. In this case the frequency of the broadcast station, controlled by the spectroscopist, is tuned until the receiving molecule responds. Clearly, if the frequency of response can be interpreted as accurately as it can be measured, very detailed and accurate information about the molecule should be obtained.

Any exact interpretation must be done by means of quantum mechanics. The fact that a particular frequency of rotation, rather than all possible frequencies, is observed is already a clear violation of what would be expected from classical mechanics. Quantum mechanics characteristically allows only certain discrete energies or frequencies to occur. All the behavior of the electronic equipment and microwaves may be very successfully treated by classical physics, but in the interaction between microwaves and isolated gaseous molecules, the characteristic phenomena of quantum mechanics appear. One of the most fundamental results of quantum mechanics states that angular momentum always occurs as some integer J times the fundamental unit $\dfrac{h}{2\pi}$, where h is the famous Planck's constant. The angular momentum may be written as the angular velocity times the moment of inertia I, or $2\pi\,\nu I$, where ν is the frequency. Hence the frequencies of rotation for a simple molecule such as sodium chloride are given by the formula $\nu = \dfrac{Jh}{4\pi^2 I}$. The integer J may be 1, 2, 3, or any other value, so that the possible frequencies of rotation of sodium chloride or similar molecules are given by Figure 5. The rotational spectrum of a simple molecule is then a series of frequencies which are harmonics of the lowest frequency $\dfrac{h}{4\pi^2 I}$. For molecules of medium or small size, the moments of inertia are such that the lowest frequency falls at wavelengths between 10 centimeters and one centimeter. This type of rotational spectrum is frequently seen in the infrared region where it is superimposed on vibrational frequencies. In many cases, however, infrared spectroscopy has insufficient resolution to separate the individual lines of such a series. They more or less run together, forming a band and producing the so-called band spectrum characteristic of the infrared region.

Let us now compare what is seen by microwaves with what is seen by infrared spectroscopy. In the infrared region, essentially the entire rotational spectrum superimposed on vibrational frequencies is seen as a more or less continuous band. Individual lines may not be resolved. In the microwave region, usually one individual rotational line is seen and other rotational lines are so far away on the microwave scale that a particular microwave oscillator cannot be sufficiently changed in frequency to allow them to be observed. Infrared gives a good general view of the spectrum. By comparison, microwave spectroscopy is like a very high-powered, high-resolution microscope which can see in great detail a very small region of the spectrum but, like a microscope of high magnification, cannot readily go from one small part of the spectrum to another. The difference in magnification between the two techniques—or more important, the typical difference in resolution—is approximately a factor of 10,000.

One might wonder if any detail is present to be seen with this gain in resolution, or whether an accurate measurement of single rotational lines is all that can be obtained. It will be shown below that, as is usual in the discovery of nature, a closer and finer examination discloses always finer details. The discovery of high-resolution microwave spectroscopy immediately after the war was in fact much like the sudden discovery of a new type of microscope with resolution greater by 10,000. Many details of spectra which were known to exist but were somewhat blurred by the lower resolution of infrared and optical spectroscopy could suddenly be seen clearly, and other less expected effects have been brought to light.

Before proceeding to a detailed examination of the effects and spectra observed, let us indicate just how these spectra are detected and the nature of the microwave equipment. Some of the devices developed for the 1.25-centimeter radar and now pressed into use by microwave spectroscopists are shown in Figures 6 and 7. Microwaves are generated by the motion of electrons under the influence of electric and/or magnetic fields. Figure 6 shows two types of microwave generators known as klystrons. The klystrons are rather small devices approximately the size of an ordinary radio vacuum tube. When proper electrical voltages are connected to the pins on the socket of these tubes, microwaves are generated and emerge

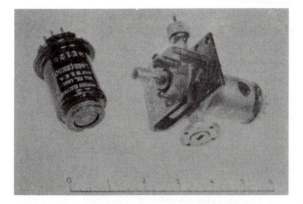

Figure 6. Two klystron tubes which generate microwaves of about 1 cm. wavelength. Scale is in inches.

Figure 7. Waveguides and a T-junction for transmitting micro- waves of about 1 cm. wavelength. A crystal detector mount is shown on the right. Scale is in inches.

from a window in the tube. In the case of the smallest klystron in Figure 6, this window is at the end of the tube and can be seen in the photograph. From the outside, these devices appear to he simple. However, their nor- mal cost in 1950 dollars of about \$350 each[*] gives a somewhat better idea of the complexity and exactness of the internal structure of the tube. Al- though building microwave oscillators is not easy, now that these tubes have been designed and are on the market it is a very simple matter to at- tach the proper voltages and obtain microwaves. Various standard types can provide almost any wavelength down to 5 millimeters, although good operation of the highest-frequency tubes is still an interesting gamble.

After the microwaves have been produced in an oscillator, one must

---

[*] At one time plus a luxury tax of 20 per cent.

consider their propagation. Ordinary long-wavelength electrical waves or currents require two conductors for their propagation. Optical waves, on the other hand, are usually propagated through the open air without the use of any conductors. Microwaves are generally intermediate in behavior between the longer electrical waves and the optical waves. They can be propagated along two conductors or in the open atmosphere, but more typically they are transmitted along one single conductor of the type shown in Figure 7. These are hollow pipes of conducting material. The microwaves can travel on the inside of these pipes, but in order to do so they must be short enough to "fit into" the pipe. More precisely, their wavelength must be comparable with the dimensions of the cross section of this pipe. Once the waves are introduced into the conductor they may be guided around through tortuous curves or bends as shown by the bent tubing in Figure 7. Perhaps for this reason such tubing is usually called a waveguide. The waveguide may also be split into a T-shape, as is shown in Figure 7. The microwaves entering such a junction through one waveguide will split. Part goes into one arm of the T and part into the other arm. These waves may be piped around very much like water, and in fact arrangements of waveguides are frequently called microwave plumbing.

After generation and transmission, the remaining essential part of any microwave system is detection. Detection is accomplished by a silicon crystal rectifier essentially like the crystal detectors of the early crystal radio receivers. These contain a piece of silicon and a very fine wire called a "cat whisker" which makes contact with the silicon. The detector is usually inserted into a waveguide "crystal holder," such as that shown at the right in Figure 7. Microwaves reaching a crystal detector produce a voltage in the detector proportional to their intensity, and this voltage is carried by the cable shown attached to the detector in Figure 7.

Microwave Spectrographs

A simple but complete microwave spectrometer is shown schematically in Figure 8. The waves are generated in the vacuum tube on the left, transmitted through a waveguide containing gas, and received by the detector on the right. The gas is normally at very low pressure—approximately 10^{-5} atmospheres—and is isolated from the atmosphere by means of mica windows at the two ends of the absorption cell. The frequency of the microwaves traversing the gas can be varied by altering the voltages applied to the klystron generator. If the gas absorbs some particular fre-

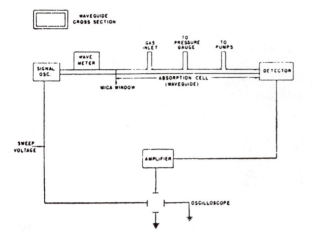

Figure 8. Schematic plan of a simple microwave spectrometer.

quency, the power at that frequency transmitted through the absorption cell is diminished as indicated by a reduced voltage produced by the crystal detector. The reduction of voltage may be observed on a cathode ray oscilloscope or a recording galvanometer. If an oscilloscope is used, as indicated in the figure, the voltage which is applied to the klystron, in order to change its frequency, is also applied to the horizontal sweep of the oscilloscope. Hence, the horizontal axis corresponds to the frequency variation. The output voltage of the detector is applied to the vertical sweep of the oscilloscope so that the vertical axis indicates the power reaching the detector, and any sudden decrease of this power corresponds to an absorption. The oscilloscope then shows directly a plot of the gas spectrum in the frequency region over which the klystron is being swept.

A picture of such an oscilloscope trace is shown in Figure 9; the three dips or valleys correspond to three different frequencies of response of the gas. This type of spectrometer is extremely simple and can be used for many spectra, but for the weaker spectra it is not sufficiently sensitive. For an absorption cell one meter long, the strongest known microwave absorption would reduce the power transmitted by only 10 per cent. Some of the weaker known absorptions would reduce this power by only one part in 10^7. Since the power output of the klystron can easily vary by a small percentage, it may be expected that such small changes due to gas absorption cannot easily be discriminated from other possible causes of power change. Fortunately there are other, more sensitive types of spectrometers.

Probably the easiest method for obtaining increased sensitivity is to utilize what is known as the Stark effect. When rotating molecules are

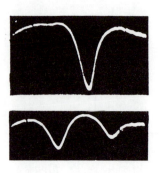

Figure 9. Oscilloscope traces of absorption lines of microwave frequencies. Upper trace: absorption line due to OCS. Lower trace: same line split into two components by Stark effect. (From Dakin, Good, and Coles, Phys. Rev., 70, 560, 1946).

placed in an electric field, the field exerts a torque on the molecule because of its interaction with the molecular dipole. These torques can twist the molecule and affect to some extent its rotational motion. The effect on spectra of an electric field was discovered a number of years ago by the German physicist Stark, who observed a splitting of atomic hydrogen lines when hydrogen was in a very strong electric field, each split component corresponding to one way the electric field could affect the electron's motion in the hydrogen atom. The Stark effect in molecules could be predicted, but the phenomenon could not clearly be observed by infrared or optical spectroscopy because of inadequate resolution. With microwave techniques, however, it is very easy to see the effect of an electric field on a molecular spectrum. The individual spectral lines change frequency or split into a number of separate lines. Detectable effects can be seen at times with electric fields as weak as a few volts per centimeter, whereas Professor Stark used a field strength as large as 100,000 volts per centimeter to see an effect in the optical hydrogen spectrum.

One method of producing Stark effects is to insert a metallic plate in the center of the rectangular waveguide and parallel to its broadest side. Such a septum does not appreciably disturb the propagation of the microwaves. The plate is insulated from the outer part of the waveguide; a voltage difference between it and the waveguide may hence be maintained. This produces an electric field in the region where molecular absorption may occur. The effect of such a field on one particular microwave absorption line is shown in Figure 9. The line is split, and the absorption frequency of each component is different from the original frequency before the electric field was applied.

A spectrometer for observing the Stark effect and also for using it to obtain very sensitive detection is shown schematically in Figure 10. The

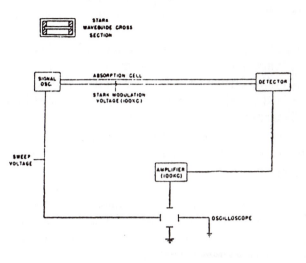

Figure 10. Schematic plan of a simple microwave spectrometer using Stark effects.

waveguide in which the absorbing gas is placed has a central conducting septum, as is shown in the cross-section at the top of the figure, and between this plate and the surrounding waveguide an electrical voltage may be introduced by an electrical connection which is brought out through a hole in the waveguide. As the voltage is varied, an absorption line may be observed to split and change frequency as indicated in Figure 9.

For sensitive detection the electric field is used to vary the molecular absorption at a particular frequency. In principle the technique is much like flushing a bird or making it move in order to see it. If microwaves of such a frequency that they are absorbed by the gas are passed through the waveguide, they may be absorbed during the time that zero field is present. However, when the electric field is large the frequency of response changes and absorption no longer occurs. Since the absorption can be made to occur or to disappear by applying the electric field, it can be more readily distinguished from other types of interfering variation in the power received by the crystal. Usually the absorption line is wiggled back and forth in frequency very rapidly by applying an alternating electric field at, for example, 100 kilocycles per second. The absorption then appears and disappears at the crystal detector exactly 100,000 times per second. The signal produced by the detector is sent to an amplifier which amplifies only signals very near 100 kilocycles per second. Since other variations in detected power are unlikely to occur at exactly this frequency, this technique allows a high degree of discrimination between genuine lines and spurious effects. The amplifier signal may be rectified or converted to a slowly varying voltage which is again used to deflect the beam of an oscilloscope. Hence, a graph of the absorption response appears on the oscilloscope as before.

With this technique extremely small absorptions may be observed. A change in power of one part in 10^7 in a distance of a few feet has been detected. This corresponds to an amount of absorption per centimeter (or absorption coefficient) of one part in 10^{-9}, and would decrease the intensity of the waves by a factor of only two in a distance of about 5000 miles. Although such a sensitive spectrometer allows measurement of extremely weak absorptions, there are many absorption lines of molecules which are still weaker, and continued effort is being made to develop still more sensitive devices.

The great usefulness of optical, infrared, and X-ray spectroscopy in general chemical analysis is well known. In time, a comparable use of microwaves for chemical analysis may also be expected. Each of these major spectroscopic regions of course involves special techniques and has particular advantages and disadvantages. The primary advantages of microwave spectroscopy derive from its extremely high resolution, high sensitivity, and the use of electronic techniques. Because very low-pressure gas is used in the absorption cell and very small absorption coefficients can be measured, a complete spectrum may be obtained from an extremely small quantity of material. In the case of ammonia—one of the stronger microwave absorbers—a spectrum may be detected with quantities as small as 10^{-12} gram. This assumes, of course, that adequate methods of handling such a small quantity of ammonia gas are developed.

Because of the high resolution of microwave spectroscopy, the absorption lines of one substance will only very rarely overlap the absorption lines of another substance. Fifty or a hundred different materials might be mixed in a given gas and the absorption lines of each individually seen and identified by a microwave spectrometer. Electronic techniques used in a microwave spectrometer allow very rapid observation of absorption lines, and hence make the method applicable to the study and control of fast reactions or rapidly changing conditions. Observation of an absorption line takes only the length of time required for one sweep on the oscilloscope; in typical cases, this is about $\frac{1}{10}$ second, but it may be as short as 10^{-4} second for a strong absorption.

The main disadvantage of microwave spectroscopy for chemical analysis lies in the fact that only gases can be studied. A high vapor pressure is certainly not needed since a gas pressure of only 10^{-2} millimeter of mercury is used. Hence, many substances ordinarily considered liquids or solids would have sufficient vapor pressure. However, there are many other substances which do not have even this much vapor pressure and therefore are not appropriate for microwave study. Sodium chloride, which has been used as an example, has insufficient vapor pressure at ordinary temperatures for microwave work, but can be studied at a temperature near 700°C. The gas must also have an electrical dipole moment, that is, it must contain such a distribution of positive and negative electric charges that a microwave electric field can twist the molecule by acting on these charges. Such common symmetric molecules as, for example, nitrogen, hydrogen, and carbon dioxide have no dipole moment and therefore cannot be studied by this method.

Although routine application of microwave spectrometers appears likely for the future, present instruments are still specialized and unreliable apparatuses, unsuited for routine analytical use. The microwave oscillators, which are now being made in small quantities, are unfortunately expensive, finicky, and short-lived.

Examining the Structure of Molecules and Nuclei

Microwave spectroscopy has so far been used primarily for the determination of fundamental physical and chemical properties, and it is for this purpose that present spectrometers have been designed. As noted above, the frequency of rotation of a simple molecule is inversely proportional to the moment of inertia, and is given by $v = \dfrac{Jh}{4\pi^2 I}$. Since the frequency can be measured to an accuracy of one part in 10^7 if necessary, the accuracy of determination of the moment of inertia is limited only by the small uncertainty in the value of Planck's constant h. The moment of inertia depends of course upon the masses of the atoms which make up the molecule and upon the distances between them. If the masses are known, a very accurate measurement of the internuclear or interatomic distances is obtained.

The internuclear distance is not the only quantity which may be measured. An examination of what we have been calling a single rotational line usually reveals under high resolution a complex structure of many lines—in some eases several hundred—and interpretation of this fine or hyperfine structure yields many details of the structure of the molecule responsible for the lines and of the nuclei which it contains.

One reason a rotational line is split into more than one is the presence of isotopes. Sodium chloride (NaCl), for example, may be made of either one of two common chlorine nuclei or isotopes of atomic weight 35 or 37. Sodium chloride molecules with different Cl isotopes have different moments of inertia and hence different rotational frequencies. Measurement of the ratio of these frequencies is one of the most accurate methods of comparing the masses of the two Cl isotopes. Comparison of intensities can also reveal the relative abundance of the isotopes.

The next largest source of splitting is molecular vibrations. The sodium chloride molecule can vibrate with various possible amplitudes allowed by quantum mechanics. Each of the different possible vibrational motions then has a slightly different average moment of inertia and therefore a slightly different rotational frequency. A study of these different frequencies gives information about interatomic forces within the molecule.

It has already been noted that an electric field can split a microwave absorption line into several components. The magnitude of this effect depends on the strength of the electric field and also on the size of the electric dipole moment of the molecule—that is, the distribution of charges in the molecule which allows the electric field to affect its motion. Hence, a measurement of the amount of splitting produced by a given electric field determines the molecular dipole moment. This method is more accurate in most cases than other techniques for measuring dipole moments. It also has the great advantage of determining the molecular dipole moment for molecules in a particular rotational and vibrational state, and has detected variations in the dipole moment with varying amounts of vibration.

Similarly, a molecule may be subjected to a magnetic field. This produces another type of splitting called the Zeeman effect after the Dutch physicist who discovered similar effects in atomic spectra. The magnitude of Zeeman-splitting makes it possible to determine the magnetic fields associated with the molecule as a result of its rotation. For a few molecules, O_2 for example, these magnetic effects are large enough so that even the small magnetic field of the earth produces a splitting which may be seen by microwave spectroscopy.

The source of splitting of rotational spectra which has perhaps been of most interest to physicists is known as hyperfine structure, because it is usually very small. The reason for this interest is that hyperfine structure is produced by various kinds of nuclear effects, and its study therefore yields information about the structure of nuclei present in the molecule.

For most purposes, a nucleus may be considered as a small round sphere with essentially no structure, but with a total positive charge just equal to its atomic number. However, nuclei are not necessarily spherical. A typical nucleus spins about an axis with an angular momentum which is called its spin and which is some integer or half-integer times $\dfrac{h}{2\pi}$. With respect to the spin axis, the nucleus may be elongated like a cigar or it

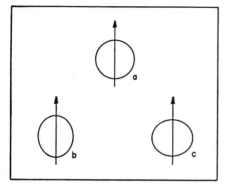

Figure 11. The shapes of atomic nuclei. a. A spherical nucleus. b. Nucleus with positive quadrupole moment (elongated). c. Nucleus with negative quadrupole moment (flattened). Arrows represent axis of nucleus and direction of its angular momentum or spin.

may be flat like a pancake, as indicated in Figure 11. An elongated nucleus is said to have a positive quadrupole moment, and a flat nucleus a negative quadruple moment. The quadrupole moment is simply a measure of the deviation of the nucleus from the spherical shape.

Consider now such a nucleus as part of a molecule. The chlorine nucleus, for example, is a flattened nucleus with negative quadruple moment and constitutes part of the molecule ClCN. This molecule is linear, that is, the atoms are arranged in a straight line. Because of the electrons and nuclear charges, rather strong electric fields exist in the molecule, and these interact with the Cl nucleus and with its quadrupole moment. According to the laws of quantum mechanics, the nucleus can take on only certain specific orientations in the molecule; these orientations are determined by the spin of the nucleus and the angular momentum of the molecule. Because of the nonspherical nuclear shape, there is an electrostatic interaction between the nucleus and the molecule which exerts a torque on the molecule and hence slightly modifies its rotation in a way that depends on the particular nuclear orientation. This disturbance results in a splitting of molecular rotational frequencies into several possible frequencies.

An observed rotational transition for ClCN is shown in Figure 12. The two isotopes of chlorine produce two different rotational lines which are very far separated, the particular ones shown being due to chlorine 35. The various vibrational effects also split this line, and their splitting is again so large that it is beyond the limits of the figure. At the top of this figure there are three different responses, or three lines, due to the several possible orientations of the chlorine 35 nucleus in the ClCN molecule. The distance between these various lines or responses is about 20 megacycles or something less than 10^{-3} wave number. The lines are still seen, however, at rather low resolution for microwave spectroscopy. If the gas pressure is decreased so that the resolution is improved and the whole

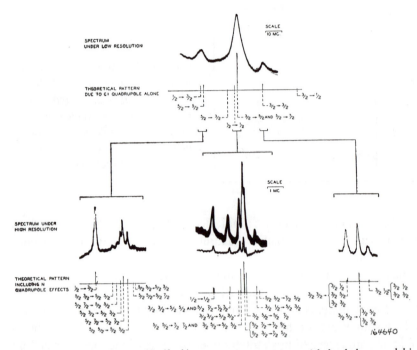

Figure 12. Spectrum of $Cl^{35}C^{12}N^{14}$. J = 1, 2 transition with both low and high dispersion and comparison with theoretical pattern. Frequency approximately 23,885 megacycles per second.

spectrum magnified more, an even finer splitting appears, as shown in the lower part of Figure 12. Each of the three components splits into several separate frequencies. This further splitting is primarily due to the quadrupole moment of nitrogen. Nitrogen 14 is an elongated nucleus (positive quadrupole moment), and since it is smaller than the chlorine nucleus, a still finer hyperfine structure is produced. The remaining nucleus, that of carbon, is exactly spherical and shows no effect of this type.

One of the gratifying and characteristic aspects of science is the quantitative theoretical prediction of the results of experimental measurement. The satisfaction of very accurate agreement between theory and experiment is prominent in microwave spectroscopy With the proper choice of a few parameters, theoretical predictions can be made to agree with experimental measurements with an accuracy as great as one part in 10^6. Theoretical calculations of the spectrum to be expected from ClCN are shown in Figure 12 immediately below the experimental observation. Both the positions of the lines and their intensities are predicted, and they can be seen to agree with the experimental values extremely well.

Even though the agreement between the expected hyperfine structure and that found experimentally is very good, there are in fact some discrepancies as large as one part in a few million. These discrepancies are large enough to suggest the presence of some new type of nuclear effect. They may be due to a polarization of the nucleus by molecular electrons—an effect analogous to the production of tides by the moon; an electron moving in orbits in the molecule is in the place of the moon, and the nucleus represents the earth. The electron, because of its electric charge, exerts large forces on the nucleus, thereby producing a tide of positive charge which follows the electron around in its orbit. The resulting distortion of the nuclear shape should give a small variation in the molecular rotational frequencies. This effect has not yet been established but, according to calculation, may be large enough to be observed with microwave techniques.

Still another source of hyperfine structure is associated with the magnetic moment of the nucleus. This interacts with the molecule since the rotating molecule involves some magnetic as well as electrostatic fields. In addition, the magnetic moments of two nuclei in the same molecule may interact and produce hyperfine effects.

Figure 12 shows that in theory several of the observed lines are multiple lines, and if still higher resolution could be obtained still further lines would be observed. This is a good illustration of the general observation that every closer examination of nature reveals more complications and finer details. Undoubtedly there remain to be discovered still finer details that are as yet unsuspected.

Atomic Clocks

We have been discussing the many types of fundamental information that have been obtained as a result of the technological development of radar. The historical development in this case is just the opposite of what most "pure" scientists like to regard as typical. They would argue that in most cases pure science develops principles and ideas, which are then applied to technology by other scientists and engineers and incorporated in equipment and devices. But in the case of microwave spectroscopy, the reverse has occurred. The equipment and devices were developed first, and pure science owes a considerable debt to technology. There is, however, every reason to believe that microwave spectroscopy will in time be able to repay this debt. It seems likely that work in this field wil lead to the use of groups of molecules to generate, detect, attenuate, of filter microwaves. These groups of molecules can in many ways do at high frequencies what ordinary circuit elements such as resistors, condensers, and inductances do at lower frequencies. One possible contribution to technology which is already being developed is the so-called atomic clock.

Any clock is based on the periodic motion of some device. The ordinary pendulum clock, of course, depends for its accuracy on the periodic motion of its pendulum. Our present most accurate clock, and therefore our time standard, is an astronomical one depending on the periodic rotation of the earth and its constancy. However, an intercomparison of the earth's rotation and its orbital motion has shown that the earth's rotation does in fact vary from time to time as much as about one part in 10^8, or one second every few years. While this is not serious for ordinary affairs such as getting up in the morning, even better accuracies are desirable for certain types of astronomical and physical measurements and for some radio and navigational uses. The reasons for the variation in the rate of the earth's rotation are not well known. A regular seasonal variation, however, has recently been explained in terms of the seasonal motion of the winds. During one part of the year these winds, blowing against the long ranges of mountains in North and South America, tend to slow up the earth's rotation. During other times of the year the predominant winds blow in the opposite direction and speed up the rotation again. Causes of other variations are more obscure but may be associated with shifts in icecaps or other earth masses. All this means that the earth is not a really reliable clock—at least not for accuracies better than one part in 10^8.

Atomic and molecular motions are also periodic and hence in principle could be used as the basis for a clock. The microwave frequencies of the ammonia molecule, for example, provide excellent frequency standards, and so-called atomic clocks are being designed to use ammonia molecules essentially as a pendulum. The ammonia molecule is constructed like a pyramid with three hydrogens at the base and nitrogen at the apex. This molecule can invert itself like an umbrella turning inside out, so that the pyramid points in just the opposite direction. The inversion takes place periodically and regularly about 24 billion times per second. This rate is very slow for a molecular vibration, and the ammonia microwave spectrum is unusual in that it is due to vibration rather than rotation. The vibration has just the frequency required to respond to radiation of the ill-fated 1.25-centimeter radar.

The ammonia molecule has many properties that make it ideal for time standards. So far as is known, the inversion frequency of an isolated ammonia molecule is completely constant, because it depends only on the

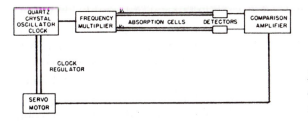

Figure 13. Block
diagram of an "atomic
clock."

universal physical constants and the nature of the nitrogen and hydrogen atoms. The frequency cannot be changed since no small changes within the molecule can occur. If the molecule is varied the smallest possible amount, by removing one electron, for example, it is almost completely disrupted and changed. This is rather different from the situation in regard to the earth, where innumerable small changes can occur. Another great advantage in using a molecular or atomic motion as a time standard is that this standard can be exactly reproduced anywhere and any number of times. Ammonia and its frequencies are the same no matter when or where the ammonia is produced. However, there remains the problem of attaching some sort of dial to the ammonia molecule so that one can actually tell time. A good atomic clock has not been possible because high-frequency electronic techniques have overlapped the frequencies of favorable atomic and molecular motions. However, electronic circuit oscillations can be made to synchronize with the molecular motions and can be controlled by them. These electronic circuits can in turn control a standard clock of the usual type.

One system which would synchronize an electronic clock with the motions of ammonia molecules is indicated diagrammatically in Figure 13. This scheme uses a vibrating quartz crystal, which in itself is an excellent clock although it does drift in frequency over a long period of time. The oscillation frequency of this quartz crystal is multiplied by using its harmonics, and from them two frequencies very close to the ammonia line are selected, one slightly lower than the ammonia frequency and the other slightly higher. Each one is sent down a separate absorption cell. If they lie symmetrically on either side of the maximum absorption due to ammonia, then these two frequencies will be equally absorbed as they pass down the two arms shown in the figure, and the amount of power received by the two detecting crystals will be equal. If the quartz-crystal clock becomes too slow, then the high-frequency harmonic is more strongly absorbed than the lower-frequency harmonic. The signal reaching the two detectors is then unequal, and the amplifier sends a command to the servo motor to adjust the quartz-crystal clock so that it runs somewhat faster. Adjustment continues until the amplifier receives and sends out zero signals. If the quartz-crystal clock becomes fast, the correction is in the opposite direction. Thus the quartz-crystal clock is continuously synchronized with the ammonia resonance frequency. This particular system is theoretically a simple one. However, it has some technical difficulties. The two arms or two waveguides would have to be very closely identical or balanced, as well as the two detectors. Many other possible schemes of synchronizing electronic oscillators with ammonia frequencies are being studied and developed. One of these systems is essentially like the one shown in Figure 13, but to avoid the difficulty mentioned, both frequencies are transmitted down the same waveguide and received by the same crystal detector.

It is difficult to predict how accurate atomic clocks will ultimately be. Potentially, the accuracy should be at least as high as one part in 10^{12}. However, a large amount of engineering effort and ingenuity must be invested to achieve good accuracy. It is engineering difficulties rather than the potentialities of the ammonia molecule which will probably limit the ultimate accuracy of the ammonia atomic clock. The National Bureau of Standards already has an ammonia clock with an accuracy of about one part in 10^8. Perhaps within the next few years, and after sufficient engineering ingenuity has been applied, atomic clocks of super accuracy will be developed.

More accurate clocks mean more accurate measurement of molecular resonances and the possibility of a wide range of precise and interesting experiments in pure physics and astronomy. Thus pure and applied science may be expected to continue their closely interdependent progress.

Acknowledgment

Much of the work mentioned here has been supported jointly by the Signal Corps and the Office of Naval Research.

Postscript

Microwave spectroscopy was discussed above as the new and rapidly developing field it was in its early years. It has, as expected, matured and provided further determinations of the structure and behavior of a large number of molecules. It has also been extended to include the study of many unstable and transient species, use of molecular beams, and parallel work at far infrared wavelengths. Still more strikingly, but quite in keeping with the nature of scientific development, completely new fields have

sprouted from it and some have themselves grown very large. They include quantum electrons or the maser, which was just barely suggested
above in the discussion of molecular circuit elements, and the maser's
growth into the wonderful field of lasers. Another vigorous extension of
microwave spectroscopy is into interstellar clouds and circumstellar materials, with the discovery of the microwave spectra of many complex
molecules in interstellar space and even of astronomical masers. This
work has opened up important and exciting new areas of astronomy. Masers, lasers, and microwave studies of interstellar molecules are discussed
more fully in the two following essays.

Masers

E lectronics is a young field, and perhaps subject to even more rapid change than most fields of science or technology. One visible and likely stimulus for change in the character of electronics during the next decade is the maser, which has been taking form and gradually emerging from research laboratories during the ten years following its first emergence in 1954. The maser is a device based on a new type of amplification, with properties and potentialities almost qualitatively different from what has previously been available. It plays a key role in a new field sometimes called "quantum electronics," which is concerned with the production and control of electromagnetic waves by their interaction with electrons bound in atoms or molecules, instead of those flowing through a vacuum as in a vacuum tube, or through a semiconductor as in a transistor. This takes us into a world that was once of interest only to the fundamental sciences—the structure of atoms and molecules, spectroscopy, and quantum mechanics.

The phenomenon of amplification itself may be characterized as a controlled release of stored energy. In the case of the maser, or its derivative now known as the laser, energy is stored in the excitation of individual atoms and molecules. It is released by subjecting the molecules to electromagnetic waves of just the right frequency which stimulate the molecules to emit their excess energy. The released energy is given up as electromagnetic radiation to the stimulating wave, which is thus amplified. Hence the name maser, which was originally an acronym for *m*icrowave *a*mplification by *s*timulated *e*mission of *r*adiation. The technique is by no means limited, however, to microwaves and has now been extended over such a wide range of frequencies that the name should perhaps be inter-

Originally published in *The Age of Electronics*, 1962 by McGraw-Hill. Reprinted with permission of McGraw-Hill.

preted as *molecular amplification by stimulated emission of radiation*, since molecules are the important amplifying element. But its application to light has generated the word laser, for *light amplification by stimulated emission of radiation*. It has also been suggested that the word "maser" has a cabalistic meaning of particular significance to laboratory directors and to government agencies, i.e. *means for acquiring support for expensive research*.

Considered as an invention, the maser epitomizes the great change that has recently come over the character of technological frontiers. It was worked out and predicted almost entirely on the basis of theoretical ideas of a rather complex and abstract nature. This is not an invention or development which could grow out of a basement workshop, or solely from the Edisonian approach of intuitive trial and error; it is rather a creature of our present scientific age which has come rather completely from modern physical theory.

The remarkable properties and versatility of this new device offer exciting possibilities to a wide variety of fields. As amplifiers of radio waves, masers are 100 to 1,000 times more sensitive than those which were previously available. We thus suddenly have amplifiers which no longer contribute significant noise and which increase markedly the possibility of long-distance communications and radar. The use of maser amplifiers in the West Ford terminals has enabled Lincoln Laboratory to establish a 50,000 bit/sec transcontinental communication link via the moon. Frederick Kappel, president of American Telephone and Telegraph, has said, "Without the maser we do not see how we could even hope for high-quality space communications."[1]

Large radio telescopes already probe somewhat farther into the universe than the most powerful optical telescopes. Because of greater sensitivity, maser amplifiers on these telescopes can increase even their range by a factor of almost 10, allowing us to approach still closer to the edges of the universe. Figure 1 shows an early maser used for radio astronomy,[2] placed at the focus of a 50-ft dish so that it can amplify the very weak radio waves picked up from outer space. This particular maser is neither small nor cheap. But the potential gain in sensitivity would allow the 50-ft telescope as great a range as a 500-ft telescope using an ordinary amplifier. Maser amplifiers on radio telescopes have produced improved measurements of the temperatures of planets, and the first measurements of the amount of hydrogen in a number of galaxies.

A quite different facet of masers is seen in the atomic clock. Not very long ago, the best clocks accumulated errors at the rate of about 1 sec every 10 years. Maser oscillators operate with such constant frequency that they can be used to control clocks which accumulate errors at a rate

Figure 1. National Research Laboratory 50-ft. radio telescope with maser installed at the prime focus. (Courtesy of National Research Laboratory.)

of 1 sec in 10,000 years. There are now hopes for maser clocks of still greater accuracy.

Another important aspect of masers, which is still rather different from the last two, is that they provide amplifiers and oscillators at frequencies which are enormously higher than those previously attainable with electronic tubes. They allow us to expand electronic-like techniques from the normal radio region into the infrared and optical regions of the electromagnetic spectrum—and probably beyond. Their making available this expanded frequency range to refined and flexible techniques has been compared in importance with opening up of the microwave region, on which so much of present-day radar and communications is based. Masers operating in the optical and infrared region give us a source of light which is many orders of magnitude more intense than that of the sun, many orders of magnitude more directive than any previous searchlight or radio beam, and with a frequency purity as good as the best electronic oscillators. Optical and infrared masers, or lasers, are beginning to be used for such diverse fields as radar, communications, microscopy, welding, and eye surgery. Research effort in this field exploded from almost zero in 1958 to an effort involving several hundred different companies and laboratories only three years later.

With these indications of the significance of masers in mind, let us now take a more leisurely and historical approach to their development. It is perhaps premature to view masers from a good historical vantage point since they are still quite new; the most active research in this field is probably still yet to come, and we are just in the midst of exploring their potentialities. However, we have at least an opportunity to discuss our subject with fresh memory of its early stages, to examine an interesting development in mid-course, and to attempt a projection toward the future. What I have to say about this development must necessarily be based on the experiences and evolving ideas of an individual in the midst of his own hobby. But perhaps it will nevertheless give some helpful perspective and useful illustrations of typical processes involved in scientific and technological progress.

All too frequently, we look back in time at some invention or discovery and contract its development into a sudden revelation to one particular man, and at one particular moment. This makes invention and scientific discovery appear spectacular and mystical, when in fact it is carried out by ordinary human beings, in small and stumbling steps. What really is marvelous is that scientific knowledge is so fruitfully accumulative. One idea or technique is added to another, making still a third one easier to understand or achieve. Soon a structure has developed which is, in fact, impressive and spectacular, even though individual steps viewed in detail are simple and ordinary.* This is clearly not meant to belittle the scientist, engineer, or inventor, but rather to emphasize the extent to which science consists of the accumulation of small, uncertain steps. If this seems belittling, it is perhaps most appropriate that I illustrate by my own field. It is also important to be aware of the extent to which science and engineering are social phenomena, evolutionary processes involving the cultural medium, the interaction between many individuals, and the hand-in-hand development of both fundamental and applied science.

In order to discuss the maser and how it amplifies electromagnetic waves, it will perhaps be helpful to review the character of these waves. As indicated in Figure 2, they include a number of familiar types of radiation (light, X rays, radio waves) and great variety. But despite this variety, they are all similar in the sense that they obey the same physical laws, differing only in wavelength or frequency. One of the least known regions lies between the microwave and infrared frequencies because the only sources of radiation in this no-man's-land of the electromagnetic spectrum are hot objects, which radiate very weakly at these relatively long wavelengths, thus allowing only primitive study of the region. At higher fre-

---

*Cf. Aristotle, "Metaphysics," A.10, 993[a] 30: "The search for truth is in one way hard and in another easy. For it is evident that no one can master it fully nor miss it wholly. But each one of us adds a little to our knowledge of nature and from all the facts assembled there arises a certain grandeur."

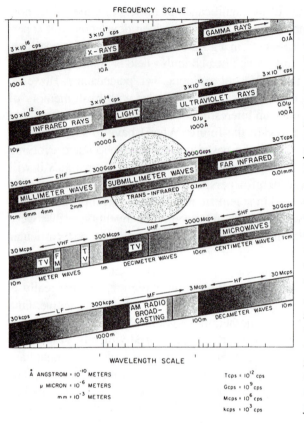

Figure 2. The spectrum of electromagnetic waves. The spectrum displayed on a logarithmic scale suffers an apparent bandwidth contraction as the frequency increases. The absolute bandwidth of any given area on the chart is as many orders of magnitude larger than the same area of a lower-frequency region as the difference in decades between the two regions. The region occupied by visible light for example, has about the same bandwidth (approximately 4×10^{14} cps) as the entire spectrum lying below it in frequency.

quencies, hot objects radiate more intensely, while at lower frequencies electronic oscillators or their harmonics are available as sources.

The maser amplifies electromagnetic waves by stimulating emission from atoms or molecules in ways well known to the physicist. The theory of atomic emission and absorption was developed as a result of intensive spectroscopic work, primarily in the optical range, over many decades Niels Bohr first suggested that light or other forms of electromagnetic radiation is emitted when an atom makes a transition between two discrete energy levels. If the atom falls from an upper level to a lower level, it emits a quantum of electromagnetic energy (Figure 3a). If it absorbs a quantum of electromagnetic energy from a wave, it must make a transition from the lower level to the upper level (Figure 3b).

In 1917, Einstein published an important paper on radiation which gave a clear exposition of the relation between the various processes by which atoms absorb and radiate electromagnetic waves. He pointed out that there must be two separate ways by which an atom may emit a quan-

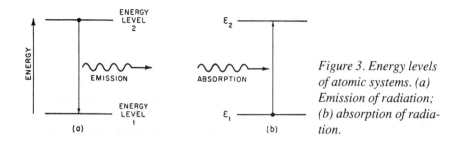

Figure 3. Energy levels
of atomic systems. (a)
Emission of radiation;
(b) absorption of radia-
tion.

tum of radiation. It may fall spontaneously from a higher to lower energy
level, producing electromagnetic waves where none was previously present. This spontaneous emission is responsible for essentially all the light
we see from the sun or from incandescent lamps. If there is present, however, an electromagnetic wave of the appropriate frequency, it may stimulate an atom in the upper state to fall to the lower state, giving up its energy to the stimulating wave. The appropriate frequency, as Bohr pointed
out, is such that the frequency multiplied by Planck's constant is equal to
the difference in energy of the two levels. These ideas are basic to spectroscopy and to the emission or absorption by atoms of electromagnetic
waves of all frequencies.

Rather recently physicists have extended spectroscopic research into
the microwave and radio regions. Perhaps Cleeton and Williams[3] in 1934
were the first to demonstrate a reasonably sharp spectroscopic resonance
in response to waves in the radio region when they examined the absorption of microwaves in ammonia. I. I. Rabi's work on radiofrequency resonances in beams of atoms came also before World War II and led to much
interesting research. But very extensive study of spectroscopy in the radio
and microwave regions was to come only later as a result of intensive development of microwave technology during the war and the familiarity
with electronics and microwaves which this gave many physicists.

A glimpse into the state of affairs by early 1945 is given by an internal
memorandum[4] of a large industrial laboratory, written to convince the research director that microwave spectroscopy was of both practical and
scientific interest.

Microwave radio has now been extended to such short wavelengths that it
has overlapped a region rich in molecular resonances, where quantum mechanical theory and spectroscopic techniques can provide aids to radio engineering. Resonant molecules may furnish a number of circuit elements of
future systems using electromagnetic waves shorter than one centimeter.
Many difficulties in manufacture of conventional circuit elements for very

short wavelengths can be obviated by using molecules, resonant elements provided by nature in great variety and with the reproducibility inherent in molecular structure.

So little work has been done in microwave spectroscopy, which we define broadly as the interaction of microwaves and matter, that the course of its future development and application is difficult to predict. Knowledge of the possibilities in this field comes largely from extrapolation of studies in the far infrared, from Rabi's work at wavelengths in the meter region, and from the few resonances of oxygen, water vapor, and ammonia which have received some attention in radar propagation investigation.

The memorandum expresses, with some foresight, high hopes for the general usefulness of molecular spectroscopy in technology. But the details were by no means clear. Devices such as the maser were certainly not imagined, as can be seen from a paragraph further along in this memorandum.

Molecular gas may be heated or otherwise excited to emit discrete bands of radiation whose frequency width may be adjusted over wide limits…. Although this application probably warrants investigation, it does not appear immediately promising because the radiation produced would be incoherent and because of intensity limitations. In most cases the molecules would dissociate before very great excitation is obtained.

The latter statement was generally accepted by scientists of the time and was based on the argument that, although molecules may produce radio-frequency waves or other types of radiation in rather narrow bands, thermodynamic reasoning showed that the intensity of this radiation could never be greater than that produced by a black body (that is, by normal, hot material), which is small at radio frequencies. The argument was sound, in general, but there were loopholes, and the lack of a clear understanding of these loopholes created a stumbling block to progress toward the maser. But there did follow an active and fruitful development of microwave spectroscopy of gases, solids, and liquids.

Microwave techniques had been carried to a remarkably high level during World War II. Their subsequent application to spectroscopy interested many scientists and furnished a great deal of valuable fundamental information, which whetted interest in extending microwave techniques to still shorter wavelengths. In 1945, the shorter microwaves produced were about 5 mm in length. My own interest, as well as that of many other spectroscopists, was much attracted to the frequency range between microwaves and the infrared, because this no-man's-land could be predicted to be rich in spectra and because it represented a challenge. Applied scientists were also strongly interested because it was an unexploited frequency range, with characteristics somewhat similar to those of the very useful microwaves.

Early in 1950, the Office of Naval Research sponsored a committee of scientists and engineers to attempt to stimulate interest in the production of waves in the millimeter and submillimeter range and to help provide and screen ideas for this purpose. The Carbide and Carbon Chemical Corporation also became interested in very short waves for stimulating specific chemical reactions and supported research toward their generation at Columbia University.

After almost two seasons of service on the ONR committee, I felt that some progress had been made, but that good sources of very short waves still seemed far in the future. In the spring of 1951, I was in Washington to attend the next meeting of the committee, and found myself sitting on a bench in Franklin Park early one morning, admiring the azaleas, then at the height of their bloom, but also wondering whether there was a real key to the production of very short electromagnetic waves. Suddenly I realized what was needed.

Clearly, for the shortest waves, we must rely on atomic and molecular resonators, since the appropriate resonators are too small for man to construct. The thermodynamic argument that molecules could not radiate more than a black body was faulty because it implied equilibrium, a state easily obviated, for example, by molecular-beam techniques, with which some of my Columbia colleagues had been working for years. Perhaps I had in the back of my mind the recent radio-frequency experiments on nonequilibrium and stimulated emission carried out by Purcell and Pound[5] at Harvard. The feedback necessary for oscillation could be provided by a cavity containing the radiation. In a few minutes I had calculated, on the usual back of an envelope, the critical condition for oscillation in terms of the number of excited molecules which must be supplied and the maximum losses allowable in the cavity.

It was very exciting to find in principle a way of producing a very short-wave oscillator, even though it did not look easy or promising, and though I thought I had been stupid to have been diverted from this approach for so long by the thermodynamic argument. Actually, much of what was involved was well known to physicists.

If all the atoms or molecules in a given sample of matter are in the upper state of two energy levels, and if these molecules are stimulated by an electromagnetic wave of the right frequency, the wave will induce some downward transitions and gain energy from the molecules. Under equilibrium conditions, there are always more molecules in the lower than in the upper state. Since electromagnetic waves induce transitions upward as well as downward, for material in equilibrium there is a net flow of molecules to the upper state, which robs the wave of energy, producing absorption. But if there are more molecules in the upper state than in the lower, the stimulated transitions produce a net flow of molecules to the lower state, delivering energy to the wave. This much of the basic principles involved was understood by any physicist thoroughly familiar with radiation theory since the time of Einstein's paper. I can remember several informal discussions among physicists shortly after the war on proposed experiments to detect stimulated emission. One finds the possibility of "negative absorption" dealt with understandingly, though briefly, in several places beginning as early as the mid-thirties.[6]

Two specific things had previously been lacking: a mechanism for obtaining net amplification, that is, an arrangement which would actually produce amplification and oscillation, and strong motivation for attempting what seemed to be a very difficult experiment.

But also, for most new ventures, the time must be ripe, and this had apparently not then been the case for masers. It was the group of physicists engaged in radio and microwave spectroscopy after the war, immersed in a combination of spectroscopy and electronics, who seem to have had the background and habit of mind required; the principal advances have come almost exclusively from this group. The three independent serious suggestions for use of stimulated emission for amplification came from three teams, each working on the microwave spectroscopy of gases—our own group at Columbia University, A. Prokhorov and N. Basov at the Lebedev Institute in Moscow, and J. Weber at the University of Maryland. The two groups that tackled the problem of making circuit losses less than the gain supplied by molecules, and hence of obtaining net gain, ended up with remarkably similar schemes, using a beam of ammonia molecules fired into a low-loss resonant cavity. In view of the wide variety of masers now known to work (some much more easily than the original system), this can hardly be a random coincidence. Here surely one detects a pervasive atmosphere of the times. The success of our own device was also much dependent on a method for producing very intense molecular beams which had just been worked out by Paul and his associates in Germany and which I had learned about through his visit to New York.

In the original ammonia-beam maser (Figure 4), a beam of ammonia molecules is emitted through a small hole into a vacuum. An arrangement of electric fields pushes and pulls on the molecules in such a way that those in the ground state are pulled out of the beam and those in the excited state focused through a hole in a resonant microwave cavity, where

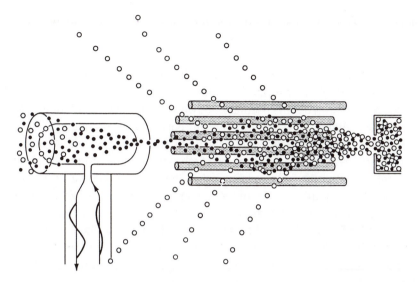

Figure 4. Ammonia-beam maser.

they can interact with the microwaves and give them their energy.

It should be remembered that, although this new type of amplifier was an interesting idea, few regarded it at the time as a "hot" topic for either physics or engineering, or gave it much priority. As a topic, its "tempera-ture" has been gradually rising ever since. Whereas new ideas on masers are now sometimes overrushed to the editor of the *Physical Review*, things were different a decade ago, as may be judged from the long delay between active discussion of masers and their appearance in standard technical journals (see Reference 6 for details of such dates). As other documentation on how uncertain our steps toward progress can be, I might mention a visit from two of my most distinguished colleagues, who urged me to stop this silly work on a molecular oscillator, which seemed very unlikely to function, because it was wasting government money. The project was being generously supported under a joint services contract ad-ministered by the Signal Corps, and I had by that time spent about two years and $50,000 on this questionable device. Their objections were a help, because the natural, and perhaps calculated, result was that I pushed on the experiment harder than ever.

In late 1953, after somewhat more than two years of work, the maser oscillated, to the triumph and credit of James P. Gordon, then a graduate student, who had risked taking on this apparently very difficult experi-ment. H. J. Zeiger, who had also helped to get the experiment under way, had by that time moved to Lincoln Laboratory.

Even at this point, the new device was not received with unmitigated enthusiasm. Another of my good friends and colleagues, after hearing a seminar on the operation of the new oscillator, advised me that he thought some of my other work was really more worthwhile. What, after all, he asked, could I do with this oscillator, now that it did work? Although I had some ready replies, I'm sure they weren't particularly convincing.

Into new research areas, our view is indeed through a glass darkly. I believe that, with a little wisdom, research will yield useful results almost as surely as summer follows winter, but their precise nature is no more foreseeable than the details of next summer's weather. We had been searching for sources of radiation in the no-man's-land between microwaves and infrared. But we had so far turned up an oscillator of remarkable stability (and hence an atomic clock)—and the realization of what an excellent noise-free amplifier the maser might become. Now there are optical and infrared masers, generally called lasers, but with all the maser's present successes, we still have none in this no-man's-land.

On the other hand, in 1951 one could already see, in a vague way, important applications for this amplifier and the possibility of a useful technology. A lecture in 1951, after some consideration of beautiful scientific results obtained from microwave spectroscopy, makes the following comments:[7]

> We have been discussing the many types of fundamental information that have been obtained as a result of the technological development of radar. The historical development in this case is just the opposite of what most pure scientists like to regard as typical. They would argue that in most cases pure science develops principles and ideas, which are then applied to technology by other scientists and engineers and incorporated in equipment and devices. But in the case of microwave spectroscopy, the reverse has occurred. The equipment and devices were developed first, and pure science owes a considerable debt to technology. There is, however, every reason believe that microwave spectroscopy will in time be able to repay this debt. It seems likely that work in this field will lead to the use of groups molecules to generate, detect, attenuate, or filter microwaves. These groups of molecules can in many ways do at high frequencies what ordinary circuit elements, such as resistors, condensers, and inductances, do at lower frequencies.

Some of the broad outlines of the future were visible, but let me disclaim any great clairvoyance by also pointing out that, a year after the maser was working, I took a sabbatical leave with the plan to examine myself and physics and to decide whether or not it would be wise to change to some other more useful type of research. I could have continued interesting enough work on the maser, but the next most useful steps were by no means clear.

Perhaps by good luck, when I settled in Paris I encountered Arnold Honig and Jean Combrisson, who were working with a paramagnetic impurity in silicon of very striking properties, which I had not previously realized could be achieved. An easy calculation showed that it could produce a maser amplifier, and I found myself again, late in 1955, in the happy state which most researchers know of losing sleep as the result of an exciting idea.

Whereas the ammonia-beam-type maser could amplify only over a very narrow range of frequencies and was practically untunable—properties which made it a very good constant-frequency oscillator—paramagnetic materials responded over a wide frequency band and were tunable, just as is needed for a good amplifier. The particular paramagnetic substance we were studying involved an electron spin which could be oriented in two directions in a magnetic field, giving it the possibility of two different energies. The separation between these two levels, and hence the frequency which might be amplified, could be changed simply by varying the magnetic field strength. Fortunately, friends at the Bell Telephone Laboratories were skilled at making the silicon necessary and were glad to supply this material when I explained our aim of making a maser amplifier in Paris. In the short time available there, we managed only to make preliminary tests and to show clearly that the material could be made to amplify.[8] Amplification was later achieved with this material by Gordon, Feher, and others at the Bell Telephone Laboratories.[9]

But such ideas were ripening in the scientific community, and very soon a somewhat different and improved approach paid off handsomely. I believe that it was in the spring of 1956 that Strandberg at M.I.T. realized independently that paramagnetic materials would make good maser amplifiers and was actively considering the use of two energy levels in such materials. A talk he gave on this possibility aroused the interest of Bloembergen at Harvard, who had been working for some time on paramagnetic phenomena. Bloembergen had an even better idea: the use of paramagnetic material with three energy levels in a suitable way would provide a more convenient and usable amplifier operating on the maser principle.[10]

The three-level solid-state maser can be understood from Figure 5. In equilibrium, the upper of the three levels is occupied by very few molecules, the next level is occupied by a somewhat larger number, and the lowest level by a still larger number (Figure 5a). If an electromagnetic wave of frequency corresponding to the energy difference between the uppermost and lowest levels is present, molecules on the lower level will

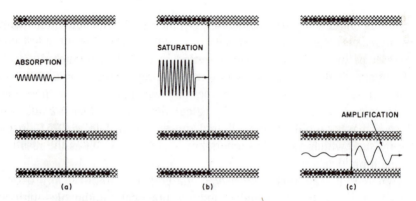

Figure 5. Three-level mode of maser operation. As suggested by Bloembergen.

make transitions to the upper level, and vice versa, until their populations are nearly equal. This decreases the population of the lower level (Figure 5b). If the decrease is sufficient, the number of molecules in the lowest level may become less than that in the second level, and maser amplification is possible by stimulated emission (Figure 5c). This system is useful partly because it can amplify continuously; i.e., by sending in continuously the electromagnetic waves which force transitions from the lowest to the uppermost state, a steady state condition in the flow of molecules can be achieved in which the central state is more densely populated than the ground state.

It is again evident that such ideas were in the air; one finds that they occurred, more or less independently, to microwave spectroscopists in several places.

Especially as a result of Bloembergen's suggested scheme, there followed a burst of activity to develop and utilize this new amplifier which theory indicated would be almost noise-free. By the fall of 1956, a number of laboratories were actively pursuing solid-state masers, and Scovil, Feher, and Seidel at the Bell Telephone Laboratories had produced the first solid-state maser.[11] Lincoln Laboratory was one of the leaders in this development; McWhorter and Meyer there were the first to measure certain of the new amplifier's characteristics; and the first application of a maser to a high-power radar was made by Kingston on Lincoln Laboratory's Millstone Hill system.

It should be noted that this burst of activity and successful use of solid-state masers would probably not have occurred if other scientific and technological developments had not produced the right setting. There was the considerable background of knowledge and understanding of paramagnetic resonances resulting from fundamental studies in microwave spectroscopy. There was also the increasing use of liquid helium brought about to a considerable extent by the Collins liquefier; a less convenient source of low temperature would have exercised considerable restraint on the use of solid-state masers. The availability of synthetic ruby, suggested as a favorable paramagnetic substance by Kikuchi at the University of Michigan, was likewise important. The development of radio astronomy and the possibility of long-range communications by means of satellites provided much of the impetus for examination and utilization of these very low noise amplifiers. The development of ferrite materials and the invention of the circulator were important ingredients in exploitation of the enormous practical gain in sensitivity afforded by maser amplifiers.

For a while those interested in masers were happily inventing new types of solid-state amplifiers, designing and engineering older types, and getting them installed on antennas for some of their initial utilization. The region between microwave frequencies and infrared frequencies was still undeveloped, and one could only hope for some good ideas to emerge for masers in this region and the result of intense activity at longer wavelengths. By 1957, however, I was impatient with waiting for good ideas and sat down to think at some length about what might be done to make masers which would operate at these short wavelengths. Progress seemed to be hindered largely by two difficulties. One was the problem of building cavities to contain the electromagnetic waves at these short wavelengths which would not also have a large number of possible modes of oscillation and which would hence be confusing and difficult to control. The other was that spectra and techniques in the region were not very well known; since no available set of molecular resonances was more than barely marginal for maser operation, the problem of struggling also with difficult techniques seemed discouraging.

The solution to this muddle was perhaps to ignore the first difficulty and jump directly into the still shorter wavelengths of the optical region where spectra and techniques were already very well known. Calculations showed that masers could be made to operate in the optical region with known spectroscopic transitions and reflecting cavities as resonators. Such masers would be very interesting in themselves. Furthermore, together with microwave masers, they could provide the basis for an assault on the no-man's-land of the far infrared from either larger or shorter wavelengths. But it was in discussions with Schawlow of the Bell Telephone Laboratories that the best ideas came to be developed.[12] He had also been actively considering optical masers, and suggested that plane parallel mirrors would be helpful. The more they were examined, the more advantages appeared for a suitable arrangement of parallel mirrors.

Consider two plane parallel mirrors between which there is a collection of atoms in excited states. A single quantum of light emitted from one atom can stimulate emission from others. If the stimulated emission is in a direction such that it is reflected back and forth between the mirrors, and if the total amount of light emitted is greater than the loss in the wave due to reflection from the mirrors, then the wave gains in intensity during repeated reflections. Ultimately, its intensity is such that it saps the energy of all atoms fed into the space and cannot grow further. This is the regime of steady oscillation.

Light can be reflected straight back and forth between the two mirrors without leaking out because of its short wavelength. Hence this system provides a resonator which takes advantage of the short wavelength rather than being too handicapped by it. Analysis showed that a proper choice of mirror dimensions and molecular properties could solve the problem of multiple modes, which had seemed so troublesome. Waves traveling in directions other than straight back and forth represent most of the multiple modes. But they are eventually reflected out of the system, and hence are not amplified. Suitable choice of dimensions of the two mirrors and of the properties of the atoms allows one to ensure that the system oscillates in only one mode and gives only a single frequency. Such an oscillator is coherent and produces as pure a frequency as the best electronic oscillators, even though it is singing at optical frequencies near 10^{15} cps.

As with maser amplification, and as with the use of three levels and paramagnetic resonance, invention of the optical maser or laser now seems in retrospect remarkably simple. The broad ideas are, in fact, very simple and represent only a small addition to what was previous understood. The details become rather more complicated, but not difficult to work out, once the appropriate train of thought is begun.

Analysis of these ideas and ways of obtaining excited atoms seemed to make it clear that optical and infrared masers would before long be made to work, and that they would have extremely interesting properties. Successful use of the spectral transitions of certain excited gaseous atoms, for which much was known, could be rather completely predicted by calculation; but the less well-known transitions in solids seemed perhaps even more promising. Excitement began to rise again and very soon there were a large number of additional suggestions for supplying excited molecules and for designs of optical and infrared masers. However, any time there is overconfidence about how clearly recognizable the most useful research projects are, I have several counterbalancing anecdotes to relate about the reception this idea received at the hands of some intelligent physicists and patent lawyers.

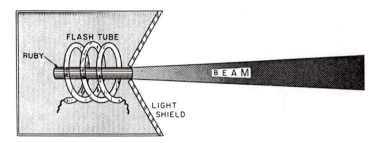

Figure 6. Ruby optical maser or laser of the type achieved by Maiman.

Maiman[13] at the Hughes Research Labs was the first to achieve experimentally maser action at optical frequencies. His method involved intense illumination of a piece of pink ruby with plane parallel faces. A group[14] at the Bell Telephone Laboratories quickly showed still more clearly than did Maiman's initial experiment that such a system (Figure 6) gives, in fact, the characteristics expected of maser oscillation. This maser operated at a rather high power level and in pulses only, because such high illumination intensity could not practically be maintained for more than a short time. The red ruby color is produced by a rather small number of impurity atoms of chromium in the hard crystal of aluminum oxide with energy levels indicated in Figure 7. Maiman realized that it was possible, with sufficient intensity of exciting light, to deplete markedly the lowest en-

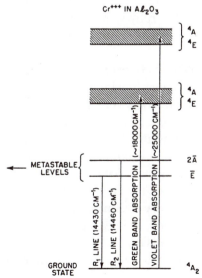

Figure 7. Energy levels of ruby.

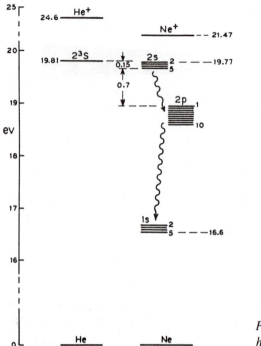

Figure 8. Energy states in the helium-neon mixture.

ergy levels of the chromium atoms. When absorbed, blue light raises these impurity atoms from the ground state to a broad spectrum of excited states. The excited atoms then drop down to an intermediate level and accumulate there in such numbers that their population exceeds the population of the depleted ground state. The ruby is then ripe for maser action. A wide variety of crystals have now been made to operate by somewhat similar schemes.

Javan, then at the Bell Telephone Laboratories but now at M.I.T., suggested[15] the use of a gaseous discharge in which helium atoms in metastable states transfer energy to neon atoms (Figure 8). The discharge excites helium atoms, allowing accumulation of metastable helium. On collision with a neon atom, a metastable helium atom transfers energy, bringing it to an upper excited state; this produces populations in the upper states larger than those in intermediate states. Again maser action can occur. Javan, Bennett, and Herriott of the Bell Laboratories made such a system operate[16] (Figure 9). Its particular charm is that it oscillates continuously, can be rather precisely controlled, and gives performance close to the best that is theoretically expected of a maser.

The optical maser is often called a laser, for *l*ight *a*mplification by

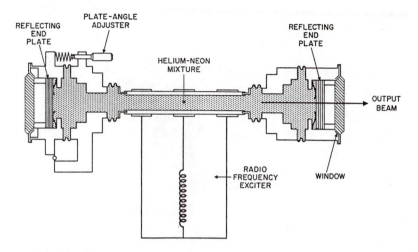

Figure 9. Optical maser as achieved by Javan et al.

*s*timulated *e*mission of *r*adiation. In order to avoid proliferation into such terms as iraser for infrared, uraser for ultraviolet, and raser for radio-frequency masers (which have all been suggested), it originally seemed to me desirable to retain maser as the general term, with the frequency range specified, as in infrared or optical maser. But laser is conveniently short, and not inappropriate for masers operating in the optical region.

By now lasers, or optical and infrared masers, are the subject of widespread and intensive research, and many types have been built. Why they were this long in being discovered is a little difficult to say. Consider the components of Javan's laser, for example. It consists only of gaseous discharge in helium and neon and two parallel mirrors. The basic quantum mechanics and spectroscopy required to understand its operation were well known thirty years ago. Why wasn't it invented at that time, in the great age of optical spectroscopy? Somehow the trend of thought had to be connected with electronics where coherent oscillators are commonplace, and to go through microwave spectroscopy, then microwave masers, and finally back to the older fields of optics, gaseous discharges, and atomic spectroscopy.

We have seen some of the many steps in the development of masers, mostly in themselves rather simple, and the scientific and technological climate which made them possible. Let us now take time for a broader view of where this has brought us. Man's pathway seems much more impressive when thus viewed broadly, either backward or forward in time, rather than in detail. Where now do we stand in the development of ma-

Figure 10. Bell Telephone Laboratories low-noise antenna-maser system. (Courtesy of Bell Telephone Laboratories.)

sers, and what further are the theoretical potentialities?

One can show theoretically that a maser amplifier should produce so little noise that it can detect an electromagnetic wave containing only one quantum of energy. It has been the custom to express the noise properties of an amplifier in terms of a "noise temperature," that is, the temperature of an object emitting heat radiation into an amplifier at a level which doubles the amplifier noise output. In these terms, good amplifiers in the microwave region have normally had noise temperatures in the range of 1000 to 2000°K. In terms of quanta of energy, this means that the amplifier can detect a minimum of about 10,000 quanta of microwave radiation. By comparison, the theoretical noise temperature of a maser amplifier in the microwave region is somewhat below 1°K, which is so low that it no longer enters into any engineering considerations. A maser amplifier has been shown experimentally to have a noise temperature less than 2°K. This is simply the best performance which can at present be measured and indicates that no amplifier noise has been detected at all. It corresponds to detection of a minimum of about 10 photons of energy in the microwave region.

With this order of amplifier sensitivity, the onus for noise troubles, very much reduced in magnitude, falls on other parts of the microwave system. Many components, whose contribution to noise could previously be ignored, now must be reconsidered and redesigned. The loss in waveguides or conductors must now be reduced, or they must be cooled to very low temperatures. Antennas must discriminate against heat radiation from the ground. The best overall system performance for an entire microwave system so far reported is a noise temperature of 20°K, demonstrated at the Bell Telephone Laboratories with a special antenna (Figure 10). This is an improvement by factor of about 100 over what could be done before the advent of masers and is possibly close to the practical limit set by the leakage of very small amounts of heat radiation into the system. Yet its performance is still two orders of magnitude poorer than that which might, in principle, be achieved with a maser amplifier, and one can hope for still further improvement.

It is also interesting to note the stimulating effect that maser amplifiers have had on the development of low-noise amplifiers of other types. Since the appearance of masers, the parametric amplifier has received increased attention and its noise temperature has been lowered to about 50°K, a ten- to twentyfold reduction. Other amplifiers have also been appreciably improved. It is likely that an important factor in bringing about these improvements was the awareness that we need no longer accept levels of noise which previously seemed irreducible.

It is partly because masers are so free of noise that maser oscillators (which are simply amplifiers with sufficient positive feedback to make them oscillate) are of such high quality and make good atomic clocks. Another reason is the fact that they oscillate only within the response width of the particular atomic or molecular resonance chosen. Some of these resonances are exceedingly sharp and stable in frequency.

There are other types of very good atomic clocks; in fact, a system using a cesium beam with no maser action at all, developed under the direction of Professor Zacharias of M.I.T., has been remarkably successful and is widely used at present. However, the clock which seems to promise the highest accuracy, or at least over short periods of time, is a direct descendant of the original ammonia maser. It is a maser oscillator using a beam of hydrogen atoms, now under development in Professor Ramsey's laboratory at Harvard.[17] An early design is shown in Figure 11. This device has not yet been well tested, but it is hoped that constancy over a long period of time of possibly 1 part in 10^{13} can be obtained.

Optical and infrared masers or lasers seem likely to have considerably more importance than other types. Optical masers are probably not generally useful as detectors of light waves, because we already have detectors which are sensitive enough to respond to only a few quanta of light. These

Figure 11. Hydrogen maser clock at Harvard. (Courtesy of Harvard University.)

are photocells or, at still shorter wavelengths, X-ray and gamma-ray counters. But as oscillators operating in the optical region, masers give us potentialities which were hardly dreamed of before their advent.

Until very recently, all sources of light have been essentially like hot, glowing bodies. The intensity of light emitted by a hot source, such as an incandescent lamp, is strictly limited by the laws of black-body radiation; that is, at a certain temperature, no source can emit more than a certain quantity of energy per square centimeter, within a certain band of wavelengths or frequencies. It is true that this energy increases as the temperature of the body increases, but the intensities which we have been able to achieve are limited by temperatures of a few tens of thousands of degrees attainable in ordinary materials. This has seemed a natural limit for so long a time that we have grown to accept it. But imagine where radio technology would have been if we were limited to the radio waves produced by heat radiation. The use of radio waves would have been sadly restricted. Fortunately, radio engineers have for a long time been blessed with coherent oscillators which could produce intensities of radiation enormously larger than those from thermal sources.

Optical and infrared masers, or lasers, have provided coherent oscillators in this new frequency range which have suddenly increased enormously the amount of energy achievable per unit bandwidth, thus allowing a technology rather more like that of electronics than that of classical optics. Intensities of these oscillators are so high that temperatures of the order of 10^{20} degrees would be required in order to reproduce them by heat radiation. This increase in energy per unit bandwidth per unit area is fundamental to the new freedom yielded by masers operating in the optical region. Concomitant with it are many interesting and useful properties.

First, consider monochromaticity, or frequency purity. The frequency of an optical maser is determined in part by the separation between the two mirrors. If this separation remains constant, the only remaining effect which can disturb the frequency of oscillation is an occasional quantum of energy spontaneously emitted by some atom in precisely the same direction as the beam. In typical masers, one can show that this will produce random fluctuations of the oscillation frequency of the order of 1/1,000, or a cycle per second. Since the light is oscillating at the rate of about 10^{15} cps, this corresponds to fractional frequency changes of 1 part in 10^{18}.

In a laser of the type invented by Javan and his associates, the essential parts consist of a gaseous discharge of helium and neon and plane parallel plates at either end of the discharge. Between these plates, a standing wave oscillates stably at a frequency near 3×10^{14} cps. If one of the mirrors is only partially silvered, some infrared light leaks through in the form of a directional beam. Experiments show that over short periods of time the oscillations represent a pure frequency to at least 1 cps. Over longer periods of time, acoustic vibrations and temperature effects move one mirror with respect to the other and produce larger slow variations of frequency. For example, if one mirror moves by 1 angstrom unit (10^{-8} cm) in distance, the frequency changes by a few tens of kilocycles, or about 1 part in 10^{10}. This provides an exceedingly sensitive detection of small motions.

The radiation emitted by such a maser is coherent not only in time, but also in space; that is, over the surface of the mirror, the light wave is oscillating almost entirely in phase. As a result, the light leaking through the mirror is highly directive and represents a much closer approximation to a plane wave than has previously been achieved. The directivity of the beam is limited only by the fundamental laws of diffraction; that is, the angular spread of the beam is given approximately by λ/D, where D is the diameter of the mirror aperture and λ is the wavelength of the radiation. For our particular case, this angular divergence of the beam is near 10^{-4} radian, about one-hundredth the divergence of sunlight or of a very good searchlight.

Moreover, still greater directivity can be achieved by using a lens to converge the beam. It will focus such a beam to a point whose size is limited only by diffraction, and hence to a diameter of about one wavelength of light; that is, the radiation from a laser can be focused so that the entire beam goes through a cross section with a diameter of about one wavelength. Such focusing cannot be achieved with incoherent light, the type we have known before. After the beam diverges again, it can be made parallel by a second lens now very much larger than the first. Its angular spread is still given by λ/D, but now D is the diameter of the larger lens. A perfect 200-in. telescope used as a lens would thus give a directivity to the entire beam which would correspond to an angular spread of about one ten-millionth of a radian. Such a beam in the atmosphere would be somewhat distorted, however, because of variations in the optical properties of the atmosphere, the same variations which produce twinkling and slight apparent motion of the stars.

Light from a good laser can be directed through an optical instrument in such a way that it will be focused on the smallest region which the optical instrument is capable of resolving. Thus, when sent backward through a microscope, the entire energy can be focused on the smallest portion of a biological cell which the microscope is capable of resolving. When focused through a telescope, the beam can be directed to a spot on the moon, for example, which is as small as the smallest region the telescope can resolve when used for normal observation.

Whereas the present gas laser produces a few milliwatts of infrared energy, the ruby optical laser produces many kilowatts of red light in a short pulse. But its beam, though highly directional under normal circumstances, falls short of the theoretical directivity by about a factor of 10, while the gaseous laser approaches very closely to the theoretical limit set by diffraction. On the other hand, it has been shown that light from the ruby laser is almost entirely coherent; hence a suitable optical system can, in principle, focus it to as small a point or as narrow a beam as in the case of the gas laser.

The field strengths produced at the focal point of a beam from an optical maser have not yet been measured, but some calculated numbers are useful as illustrations. One watt of power concentrated at the focal point would have a power density there of about 100 million watts per square centimeter, enormously higher than any other known radiant intensity. Field strengths in the electric wave at the focal point would reach about 1 million volts per centimeter for 1 watt of power, or 100 million volts per centimeter for the 10 kw which a ruby laser normally produces. It must be emphasized that this figure pertains to the field strength due to *light* waves oscillating at a frequency of about 10^{15} cps. The pressure of light at this hypothetical tiny focal point would be near 1,000 atm. Such intensities can easily tear apart atoms and molecules; laser designers have been

Material	Wavelength in Angstroms	Operation
Cr^{+++} in Al_2O_3 (ruby)	6,943 7,009 7,041	pulsed and continuous
Sm^{++} in CaF_2 (crystal)	7,082	pulsed
U^{+++} in BaF_2 (crystal)	24,000 27,000	pulsed
Nd^{+++} in $CaWO_4$ (crystal)	10,600	pulsed and continuous
Nd^{+++} in glass	10,600	pulsed
Benzophenone and naphthalene in organic glass	4,700	pulsed
Ne in He (gas discharge)	11,180 11,530 11,600 11,990 12,070	continuous

Figure 12. Characteristics of some maser systems.

comparing their achievements by stating how many razor blades one millisecond flash of the light the lasers produce will penetrate. A hole blasted through one package of 12 razor blades, including its containing carton, is about par. When focused, the beam is intense enough to evaporate the most refractory materials as well as razor-blade steel.

Many kinds of optical and infrared masers are now available (Figure 12). New types appear at the rate of about one per month, with increasing variety of wavelengths and characteristics. However, there are more in the near infrared than in the optical region; lasers oscillating at wavelengths shorter than the red are still very rare. The laser, the He-Ne system, as well as some other types, have now been made to oscillate continuously at reduced power. In pulses, lasers have produced a power of about 10 megawatts for a small fraction of a microsecond. Frequencies in a given laser have been tuned over a range of about 20 wave numbers, or 600,000 Mcps.

Intensity of light from optical masers is strong enough to produce nonlinear effects in normal optical media and hence to allow the production of harmonics of optical radiation, as demonstrated by Franken, Weinreich, and others at the University of Michigan,[18] who focused a beam of laser light on a piece of quartz, producing the second harmonic of the original light at sufficient intensity to be easily detectable. These nonlinearities also allow two frequencies to beat together, so that the sum of the two frequencies is produced.

An older and more familiar device, the photocell, also responds to light in a nonlinear fashion to give the equivalent of rectification. Since the photocell responds to the intensity of light or to the square of the field strength in the light, beams of light from two different optical masers can be sent into the same photocell, where they are mixed; the difference frequency is produced in the electron current of the photocell. This is heterodyne detection at optical frequencies.

We see from the above discussion that there are now available in the optical region coherent amplifiers and oscillators, antennas and methods of transmission, frequency multiplication and beating, rectification and heterodyne detection systems. One may also add to this list filters of various types and designs afforded by sharp optical resonances, as well as attenuators of absorbing material. Thus we now have in optical technology all the normal complement of tools and devices used by the electronics expert.

Although our attention here is primarily on electronic and applied uses of masers, their application to fundamental scientific experiments seems also very promising. Since optical masers afford new orders of precision in measurement of the two fundamental quantities, length and time, we may expect them to allow in due course reexamination of many facets of the physical world with great delicacy and precision. For example, we can now visualize the tools necessary to multiply and divide frequencies in the optical and infrared region, and to amplify the resulting new frequencies so that they can be further multiplied or divided. This encourages one to look forward to the day when radio frequencies, on which our standard of time is based, are multiplied by harmonic generation to frequencies through the infrared and into the optical region, so that we can determine on the basis of our standard of time the frequency of oscillation of optical wavelengths. Since these optical waves give us our standard of length, the product of their length and frequency immediately affords a determination of the velocity of light to a precision as great as that to which the standard of length can be defined. This procedure, in fact, obviates the necessity for separate definitions of standards of length and of time, since the velocity of light can then itself be used as an absolute standard of length.

Very preliminary engineering applications of optical and infrared masers have already been made. But it is probably more useful at this stage to examine some of the theoretical possibilities and dream about potential applications than to discuss specific and rudimentary initial work. Let us reconsider some of the classic functions of electromagnetic waves and electronics.

What about communications? Optical frequencies can completely dispel bandwidth problems. One-tenth of one per cent of the optical frequency range corresponds to the entire bandwidth so far used throughout the radio and microwave frequency range. In the optical range there is sufficient bandwidth so that all inhabitants of the earth can talk simultaneously on their own private frequency bands. But in addition, the directivity of optical beams is such that wireless communication can be maintained essentially in complete privacy and without interference from other channels of communication or noise. No one beam gets in the way of another. It is, of course, true and important that optical and infrared beams will not penetrate very far through clouds or fog. This means that for long-distance communication by electromagnetic waves of this frequency one must either use pipes or fibers at the earth's surface or go above the clouds. For communications at high altitude and in space, optical and infrared beams may be particularly advantageous since there is no cloud trouble and since the broadcasting or receiving antennas can be only a few inches in diameter and still enormously directive. Nor would the power requirements of such optical masers necessarily be very large. Work is now under way to power an optical-maser oscillator from solar radiation alone.

The distance over which these optical beams can in principle be detected is impressive. This is due in part to their directivity. For illustration, let us look into the future and imagine an optical maser which can deliver 10 kw of power continuously rather than simply in pulses as at present. Let us also assume that this light is directed by a good 200-in. telescope located above the atmosphere so that atmospheric problems produce no unwanted fluctuations in the direction of the beam. At a large distance from the solar system, its beam can look brighter than the sun. This may seem surprising, but remember that its 10 kw of power is highly directed whereas the sun's energy is radiated in all directions. Remember also that the energy in this beam is confined to a very narrow frequency range whereas the sun radiates in a broad frequency range. Relatively simple filters which allow the passage of a narrow frequency range corresponding to that produced by the laser will hence make its beam seem brighter than the sun. The laser light beam could be detected by the naked eye to distances of about $\frac{1}{10}$ light year. Obviously, then, signaling between planets in our own solar system would be easy even by use of the handiest of receivers, the human eye.

With another good telescope as a receiver, the beam could be detected up to distances of about 100 light years. This allows the possibility of interstellar signaling in the optical region. Project OSMA or SETI involves an effort to listen for radio waves which may perhaps be directed toward our earth by intelligent beings associated with other stars. Our knowledge of the technology developed by some other form of intelligent life is not sufficiently accurate to ensure that attempts to communicate with us may not be by means of optical masers rather than radio broadcasts. Perhaps we should be looking at neighboring stars with a good telescope and a high-resolution spectrometer for a spectral line much narrower than those emitted by normal atoms, that is, for light which could be produced only by something like a laser.

Although interstellar communication may be a long way off, local communication by optical beams seems immediately practical. In this area, part of the engineering problem is to learn how to modulate and de-modulate sufficiently well so that the enormous bandwidth afforded can be properly utilized. Even with much wastage, the additional load of com-munications which optical waves can carry is impressive.

What about radar? Already, light has been produced at a power level of about 10 megawatts, with a pulse length somewhat shorter than 10^{-7} sec. I see no reason why the pulse length cannot be further shortened to about 10^{-10} sec, with some corresponding increase in the power level. The range of radars using optical or infrared beams could be quite large, and the short pulses would allow a very accurate determination of distances. The directivity of such a radar beam would be several orders of magnitude better than those in current use; hence angular directions could be deter-mined with greatly increased precision. But we are faced again with learning how to utilize this new potential. In terms of the necessary engi-neering, one important question is whether we can properly take advan-tage of the extreme directivity of laser beams. Can other system compo-nents be made to match their performance, and can we achieve the necessary precise control of optical systems under high power and heating conditions?

Interference patterns produced by optical beams can be used to meas-ure large distances with still greater precision than by light radar pulses. These optical beams are sufficiently monochromatic to allow the forma-tion of interference patterns across distances of many thousands of miles. Hence, for a distant reflector, motion as small as one wavelength of light could be detected easily, assuming that the optical path to it is sufficiently stable, as it would be in space. Do we want measurements of distances to this accuracy, and what use can we make of them? Would we want to use Doppler effects at optical wavelengths to measure the velocity of ob-jects hundreds of miles away with an accuracy of a small fraction of a centimeter per second?

It is clear that there are many practical uses for accurate determination of distances closer at hand. One example is measuring the accuracy of the surface of a large radio telescope dish. A laser light beam reflected from the surface of the dish would produce interference patterns by means of which its shape could be determined to within any precision desired down to about one wavelength of light.

What about power transmission? Transmission of sizable amounts of power over long distances by light beams does not seem too much to hope for. The telescope postulated above to be in orbit above the atmosphere could concentrate the 10 kw of power of a laser beam into a spot about 6 in. in diameter 1,000 miles away. A flippant thought is that the spot is about the right size and has plenty of power to heat some spaceman's morning coffee, assuming he knows just where to put the coffee pot. Perhaps in the future satellites will be supplied with power by means of optical beams—perhaps even pushed into position by light pressure.

A good telescope placed on the moon could direct a laser beam to a particular spot on earth about the size of a football field. Is it only fantasy to ask whether this will be a future form of night illumination which can be beamed to various spots in accordance with local needs? And shall we have to face commercialization in the form of bright-light advertisements on the moon?

But back on earth there are mundane applications of intense or of monochromatic and coherent beams of light which are more immediate, and probably more important. These include etching and welding, which may be carried out with great precision and under chemically pure conditions, with surfaces touched only by light; surgery or microsurgery performed with great speed and exactness on any visible surface; increased intensity of illumination for photomicroscopy under high magnification; extremely precise measurements of distances and angles in surveying, in determination of small distortions, and in very precise control of machines.

There is little doubt that we shall see many of these applications. But precisely which ones will be most successful and most exciting, and what additional new concepts and techniques will emerge, cannot be so easily foreseen. The full nature, or even all the broad fields of application, of lasers and other quantum electronics devices are surely not yet evident. In the region between microwave and infrared frequencies, we still do not have sources of electromagnetic waves of sufficient intensity for proper scientific or technical work. But it is quite easy, now, to be convinced that masers will yield such sources. Very high frequency acoustic oscillators, operating on maserlike mechanisms, can also be expected. Masers should operate in the ultraviolet region and at still shorter wavelengths, perhaps approaching the soft X-ray region, where one can show that present ideas for masers clearly will no longer work. Coherent oscillators at wavelengths as short as those of X rays will require some new discovery or invention.

Certain aspects of the future development of the maser idea and quantum electronics can be envisaged, including a number of interesting ones which space does not allow to be discussed here. But as a humbling further reminder of the unpredictable and frequently happy surprises of research, we have only to ask what research director would have been wild enough, a decade ago, to set about the development of a very low noise amplifier, or a more accurate clock, or a very intense light source by asking his research staff to study the absorption of microwaves in gases or in paramagnetic materials at low temperatures.

For those interested in electronics, optical and infrared masers seem likely to have a particularly powerful effect in the future. They will take electrical engineers deeper into spectroscopy, very high precision optics, and the strange world of quantum mechanics. This touches on what is perhaps one of the most important aspects of the maser, recently described with some humor by an editorial writer.[19] "We propose a new name . . . to include all those new concepts and devices that allow, and even force, the scientist-engineer to break into a new field of science and technology and of thought. We propose the name gaser, coined by the well-established method from *g*rowth *a*cceleration by the *s*timulated *e*xcitation of *r*esearchers. We suggest that the maser is a gaser."

It is a pleasure to acknowledge the good counsel and considerable help given me in preparation of all phases of this article by Dr. James W. Meyer of the Lincoln Laboratory.

Postscript

The field of amplifiers and oscillators based on stimulated emission of radiation from atoms and molecules can be expected to continue vigorous growth. By 1993, it had branched out into a wide variety of fields and became a 20 billion dollar industry.

Microwave masers will probably always continue to be of importance as amplifiers and precision clocks. But they were destined to be overshadowed by the optical maser, now called the laser. Laser has become a household word and, since it is not copywritten, its popularity has even been borrowed for completely unrelated objects such as cars, knives, etc. Many of the applications of lasers were foreseen quite early in the game. However, the versatility, power and beauty of these devices are beyond any real expectations of the earlier days.

Semiconductor lasers, which seemed initially rather difficult and came along later than several other types, are now the most numerous and cheapest of all. They are widely used, for example, with compact disks, in laser printers, and along with other lasers and fiber optics are transforming the field of communications. Laser surgery has been so effective that it is being used, with various ways to project the laser light, for many operations within the human body as well as on retinas and more external regions.

Laser pulses have provided power above 10 million million watts, and an intensity above 10 million million million watts per square centimeter. Masers in astronomical objects have been discovered which broadcast the frequency of a single microwave line at a power which is a hundred times larger than all that produced by the sun. Pulses of light as short as about one hundred millionth of a millionth of a second have been produced, short enough to catch in a stationary condition the fastest of chemical reactions.

On the more gentle side, lasers have been used to sensitively manipulate small parts of biological cells, to slow and cool atoms to the lowest temperatures ever achieved, or to clean plaque from blood vessels. Lasers using single atoms have been operated, with a power of a millionth of a millionth of a millionth of a watt.

The field of quantum electronics has already penetrated and assisted many branches of scientific research, including the biological and earth sciences as well as physics and chemistry. Lasers are both working tools in a wide variety of scientific laboratories, and provide the means for new and striking studies of light itself. Quantum electronics also is the basis for new and more precise definitions of both time and length. Hydrogen masers give frequencies pure to one part in one hundred million million, and further precision is aimed for. Distances can now be defined by light waves to one part in ten thousand million, and motions as small as one ten thousandth of a millionth of a centimeter can be measured.

From the point of view of technology the field of masers and lasers is still very young, and undoubtedly this blend of optics, electronics, and atomic and molecular to produce new, surprising, and beautiful applications.

Messages from Molecules in Interstellar Space

Some time ago the great astrophysicist Sir James Jeans wrote "The normal star and the normal nebulae have little to do with life except making it impossible. Life seems to be an accident and torrential deluges of life destroying radiation the essential." Jeans reflected a common astronomical view which arose in part from the fact that hot and bright objects, the stars and bright nebulae, were most readily seen by the eye and the objects normally studied by astronomers. Jeans was also pointing out that intense ultraviolet and other radiations from stars can destroy all molecules; we exist here by virtue of a thick protecting atmosphere of gases around our planet. Otherwise, our molecules along with any others would be rapidly disassociated by solar radiation or more slowly, though inevitably, in interstellar space.

Recently, it has become clear that in addition to the very localized and special regions protected by planetary atmospheres there are other vast regions where molecules are protected from ultraviolet radiation and can exist in abundance. These are the large interstellar dust clouds, which have been long neglected because they are dark and optical light could not penetrate to reveal to astronomers what might be there. Their existence was by no means completely unknown, they seemed only less interesting than the bright regions and not easily studied. The reality of such dust clouds in interstellar space is immediately seen from Figure 1, called the "Horse Head" nebula for obvious reasons, and showing such a dust cloud illuminated by a bright nebula region. That such dust clouds are extensive in galaxies is illustrated in the picture of the great Andromeda nebula on Figure 2, where the flat spiral structure of Andromeda is interlaced with dark planes, corresponding to such dust clouds. They are even more strik-

Figure 1. The "Horse Head" nebula, with dark dust clouds illuminated by a bright nebula.

Figure 2. The Andromeda nebula, a spiral galaxy showing streaks of dust lanes.

ing in the Sombrero Galaxy of Figure 3 which, looked at more or less edge-on, shows a completely opaque disk of dust out of which protrudes a ball of starlight in the central region. Our own Galaxy is fairly rich in such dust clouds, as shown by a composite picture of the Milky Way, Figure 4. The famous cleft in the Milky Way, for example, does not represent some division between two groups of stars which form the Milky Way, but simply an intervening dust cloud which obscures some of the stars present.

New varieties of astronomy, in particular radio and infrared astronomy, have brought these clouds into prominence. Their longer wavelengths penetrate and see into the clouds, and have given fascinating information about these regions which turn out to be important and in many cases very active. We now know that in addition to dust grains they are composed of

Figure 3. The Sombrero Galaxy, which viewed almost edge-on shows dark dust clouds of the galactic disk against intense stellar illumination of the central region.

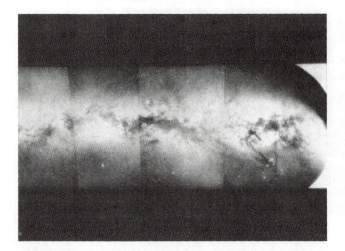

Figure 4. A mosaic of pictures of the Milky Way, where dust clouds are evident against a dense field of stars.

rather dense gas containing a rich variety of molecules, and envelop stars being generated in their interiors. They hide stellar births from us and, because stars often emit dust and gas late in life, their death is often also shrouded by dust.

Molecules within the clouds reveal their own secrets and those of the clouds largely as a result of the molecular rotational frequencies and spectra. The rotation of molecules is of course quantized in accordance with Niels Bohr's criterium, and their rotational frequencies happen to fall predominantly in the microwave and far-infrared regions. That is, they emit wavelengths which range from a few centimeters to a small fraction of a millimeter in length, and it is these wavelengths which during the last

Figure 5. Radio telescopes at the Hat Creek Observatory of the University of California.

Figure 6. The radio observatory at Onsala in Sweden.

decade have provided information about the interiors of dark interstellar clouds. Radio astronomy had been developed rapidly after the late 1940's for the study of ionized regions and atomic hydrogen in interstellar space. A few diatomic molecules in thin interstellar clouds had also been discovered as early as the late 1930's by optical astronomy, and there was the important detection of the free radical OH, found by radio telescopes in 1963. However, it was the discovery of polyatomic species in some abundance in 1968 which set off intensive activity with the already available radio telescopes in many parts of the world and which has recently revealed a very wide range of fascinating phenomena within the dark clouds.

Radio telescopes of the University of California in northern California, where NH_3 and H_2O were first detected, are shown in Figure 5. Figure 6 shows telescopes at an observatory somewhat closer to us, in Göteborg,

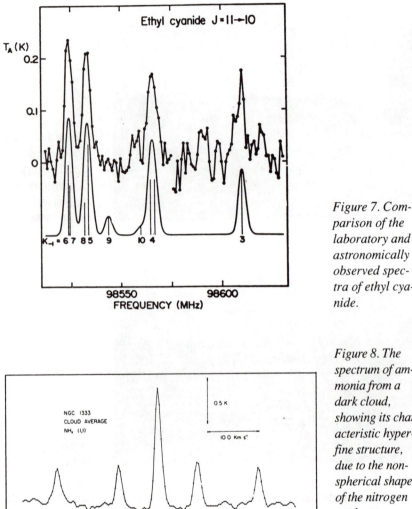

Figure 7. Comparison of the laboratory and astronomically observed spectra of ethyl cyanide.

Figure 8. The spectrum of ammonia from a dark cloud, showing its characteristic hyperfine structure, due to the nonspherical shape of the nitrogen nucleus.

Sweden, which has done much pioneering molecular work. With very sensitive receivers, for example the telescopes of Göteborg use maser amplifiers in order to enhance sensitivity, such telescopes are able to detect the weak microwaves emitted by molecules and clouds throughout our own Galaxy and even some molecules in the more distant external galaxies. One such spectrum of molecular emission is shown in Figure 7. There one sees several lines from the somewhat complex molecule ethyl cyanide. Since these spectral lines are precisely known in the laboratory, this particular pattern of frequencies provides completely definite identification of ethyl cyanide in the cloud being observed. Figure 8 shows a spectrum of ammonia from a dark cloud, with its characteristic hyperfine structure.

Fifty-two species of molecules have so far been identified. These are tabulated in Tables 1, 2, and 3, where they are grouped in several classifications. It may be immediately noted that there are a large number of organic molecules, many of them somewhat complex. These include ones which are common in biological systems or active biologically. Interstellar space appears to be as redolent with organic types as is life on Earth. While life itself may perhaps be unusual in our Galaxy, we now know that at least the basic molecules from which most biochemists believe life must be formed are produced in abundance by physical processes in interstellar space—and presumably also are already present in primitive planetary atmospheres. The tables also show many free radicals, very active and evanescent molecules, some of which have never been produced in any isolated form in the laboratory. In addition there are some very unusual species, a few of which were found first in interstellar space and yet seem to be definitely identified by their spectra. Notice particularly the long carbon chain molecules, including one with nine carbons in a row. Note also that there are no ring compounds yet discovered. Some of these have been looked for intensively and there are some technical reasons why they are difficult to find, but still we must conclude that interstellar clouds are not adept at making ring compounds. The chemistry of this collection of molecules is fascinating and still quite puzzling. It is clear that ion-molecular collisions and reactions must be important. It also seems clear that molecular reactions on the surface of dust grains play a role. However, in only a few cases can we be reasonably sure of particular chemical formation paths.

Certain common molecules appear to be missing, such as N_2, O_2, or acetylene. Undoubtedly, these are missing only because they have no dipole moment and hence no rotational spectrum with which radio astronomers might detect them. In time, some of them certainly will be detected by infrared techniques which are in the process of development.

The spectra of molecules provide marvelous probes of the conditions and history of their clouds. From them we can measure temperatures of the clouds, which are generally very cold—as low as $10°$ above absolute zero, and usually no warmer than about $70°$ absolute. As a result, most of the molecules should freeze out on any dust grains they encounter. Thus, whatever the core of the dust grain may be, it must be covered with ices of molecules, including some free radicals. Some astronomers believe a minor perturbation would therefore set off reactions in materials frozen on a dust grain and allow them to explode as a result of the chemical energy stored up in ices containing free radicals. The nature of the dust grain

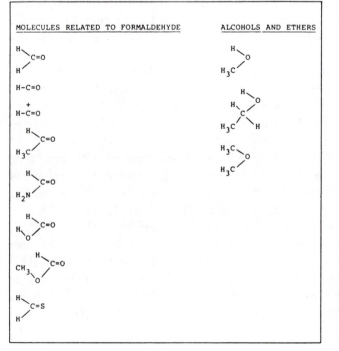

	HYDRIDES	OXIDES AND SULPHIDES	COMPLEX NITROGEN COMPOUNDS
Molecules Identified in Clouds	H_2	CO	NH_3
	OH	CS	$H_2C=N\diagup\diagdown$ H
	CH	NO	
	CH^+	HNO	$H_3C-N\diagup\diagdown$ H
	CH_4	NS	
	OH_2	COS	$H_2N\diagdown$ C=O $\diagup$ H
	SH_2	SiO	
	NH_3	SiS	H $\diagdown$ N-C N $\diagup$ H
		SO·	
		SO_2	H $\diagdown$ N=C=O $\diagup$ H
			H $\diagdown$ N=N $\diagup$ +
Common Molecules not yet Identified	CH_3	O_2	
	CH_2	N_2	
	NH	N_2O	
	NH_2	NO_2	
		CO_2	

Table 1. Oxides, hydrides and nitrogen compounds in interstellar clouds. In addition, some common molecules are listed which have not yet been found, presumably only because they have no electric dipole which allows microwave radiation.

MOLECULES RELATED TO FORMALDEHYDE

ALCOHOLS AND ETHERS

H $\diagdown$ C=O $\diagup$ H

H-C=O

H-C=O +

H $\diagdown$ C=O $\diagup$ H_3C

H $\diagdown$ C=O $\diagup$ H_2N

H $\diagdown$ C=O $\diagup$ H, O

CH_3, O $\diagdown$ C=O $\diagup$ H

H $\diagdown$ C=S $\diagup$ H

H $\diagdown$ O $\diagup$ H_3C

H $\diagdown$ H $\diagdown$ C $\diagup$ O H_3C $\diagup$ H

H_3C $\diagdown$ O $\diagup$ H_3C

Table 2. Molecules related to Formaldehyde, Alcohols, and Ethers found in interstellar clouds.

	CYANIDES	HYDROCARBONS
Molecules Identified in Clouds	CN HC≡N H–N=C H_3C–C≡N H_2N–C≡N HC≡C–C≡N HC≡C–C≡C–C≡N HC≡C–C≡C–C≡C–C≡N HC≡C–C≡C–C≡C–C≡C–C≡N H_2C 　＼ 　　C–C≡N 　／ H	C_2 C≡C–H H–C≡C–C≡C H_3C–C≡C–H
Common Molecules not yet Identified	N≡C–C≡N	N–C≡C–H H＼　　　／H 　　C=C H／　　　＼H H_3C–CH_3

Table 3. Cyanides and linear hydrocarbons found in interstellar clouds. The common molecules listed which have not yet been found have no dipole radiation, and hence cannot be detected by microwave techniques.

cores is still quite uncertain. It seems clear from the rather broad infrared spectra of dust which have been obtained that some of the dust is made of silicates similar to those contained in common rocks on Earth. Other components of the dust appear to be graphite and silicon carbide. Its exact form and composition is still much of a mystery. Presumably, much of the material in comets is composed of similar dust grains covered with frozen molecules and gathered into a massive aggregate. Thus, if we send a space probe to obtain a direct sample of one of the comets in our solar system, and this may not be very far in the future, we may have an actual sample of the material which makes up interstellar clouds. For the moment, we must use astronomical spectroscopy and other less direct measures to determine its composition, and for amorphous dust grains that is rather difficult. Fortunately, for the molecular gaseous molecules, spectroscopic results are very specific and hence we already know a great deal about the gaseous molecules.

The relative excitations of various molecular spectral lines allow a de-

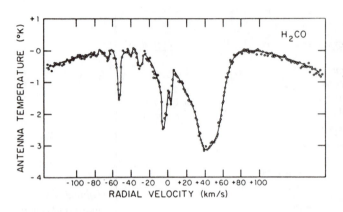

Figure 9. Lines of formaldehyde occurring at several frequencies due to the Doppler effect. Velocities of clouds in which these molecules exist and which produce the Doppler shifts are indicated as abscissae.

termination of the gas density in dust clouds. This varies from about 10 molecules per cubic centimeter, largely of hydrogen and helium, to about 10^8 molecules per cubic centimeter. On a terrestrial scale, this is an exceedingly tenuous medium, even the densest value corresponds to the best vacuum we can normally obtain. However, since the overall size of a dust cloud is large, typically in the range of one to a few hundred light years, this exceedingly low density still allows the clouds to contain an enormous amount of material. The larger molecular clouds typically have a total mass as great as that of 10^5 or 10^6 stars. There are perhaps 2000 or more such large clouds in our Galaxy, in addition to a great many smaller ones. This makes up in all a mass equivalent to one billion stars, or one percent of the total mass of our Galaxy. This material is probably the dominant state of matter in interstellar space, the remainder being still thinner clouds which are characteristically in an atomic or ionic state. The dark clouds are characteristically much more massive than the bright nebula to which we have in the past paid more attention.

In addition to molecular lines giving us information about temperatures and densities of the clouds, they provide very detailed information about cloud motions. The change in wavelength of the radiation depends on cloud velocity through the Doppler effect, giving wavelength shifts which are easily detected as is shown in Figure 9. This figure shows a single absorption line of formaldehyde split into three lines by Doppler velocity shifts, corresponding to three clouds traveling at velocities ranging from about –60 kilometers per second on the left hand side to +40 kilometers per second on the right hand side. Usually, each molecule found shows the same pattern of velocity, indicating that they coexist in the same clouds.

Another fascinating type of information that molecular spectra yields is the relative isotopic abundances in various parts of our Galaxy, providing

Isotopic Abundance Ratio

Molecular Cloud	$^{13}C^{12}C$
Sgr A	$0.048 \pm .005$
Sgr B	$0.043 \pm .006$
W33	$0.024 \pm .003$
W43	$0.016 \pm .002$
M17	$0.011 \pm .001$
Orion A	$0.013 \pm .002$
NGC 2024	$0.015 \pm .002$
W3	$0.014 \pm .002$
Average of our Galaxy	0.015
Terrestrial Abundance	0.0112

Table 4. Relative abundance of Carbon 13 and Carbon 12 isotopes found in various interstellar clouds.

Isotopic Abundance Ratio

Molecular Cloud	$^{18}O/^{17}O$
Sgr A	$3.4 \pm .4$
Sgr B	$3.2 \pm .2$
W33	$3.6 \pm .2$
M17	$3.5 \pm .3$
Orion A	$3.6 \pm .3$
NGC 2024	$3.5 \pm .2$
W3	$3.2 \pm .3$
NGC 7538	$3.5 \pm .3$
Average of our Galaxy	3.4
Terrestrial Abundance	5.5

Table 5. Relative abundances of Oxygen 18 and Oxygen 17 isotopes found in various interstellar clouds.

information about the past nuclear history of these regions. Two molecules which are otherwise identical but have different isotopes have slightly different moments of inertia. Hence they rotate at slightly different speeds, providing different wavelengths for a radio telescope to detect. Thus, by measuring the relative intensities of radiation at wavelengths corresponding to the different isotopic species, one can obtain a measure of the relative abundance of the isotopes. While there are many pitfalls to such a measurement and great care must be used, by now radio astronomers have accumulated what appears to be a fairly reliable picture of some of the isotopic abundances, as are shown in Tables 4 and 5. Table 4 lists the relative abundances found for the rare carbon isotope, Carbon 13, compared to the common carbon isotope, Carbon 12. On Earth, this abundance ratio is 0.0112. In the two clouds designated Sagittarius A and Sagittarius B, which are near the galactic center, the amount of Carbon 13 is increased by more than a factor of 3 relative to Carbon 12, the ratio being about 0.05. While various abundance ratios apply to various clouds, the overall average in our Galaxy judged from the clouds so far examined is about 0.015. It is interesting to note that this is about what would be expected from the fact that the Earth's material was frozen out of interstellar space when the Earth was formed about 5 billion years ago. The rare isotopes such as Carbon 13 are formed from the common isotopes by nuclear reactions within stars, and then thrown out into interstellar space again by explosions or stellar winds, providing material for new stars. In this way, the average abundance of the rare isotopes should increase approximately uniformly with time. Since our Galaxy is thought to be about 15 billion years old and our Earth was formed from material which was processed by successive stellar formation over the first 10 billion years, the terrestrial material should have about 10/15ths the Carbon 13 abundance of material in interstellar space which has continued to be processed for the entire 15 billion year life of the Galaxy. This is approximately what is found.

While we have commented on an average value of the $^{13}C/^{12}C$, Table 4 shows that actually this ratio appears to vary somewhat from cloud to cloud. Since about 5 billion years is required in order to change the average Carbon 13 abundance by about 30 per cent, these variations from cloud to cloud indicate both that the individual clouds must have varying rates of stellar and nuclear activity, and that the clouds must be isolated from each other over a long period of time, that is, as much as about 1 billion years. For if they were not isolated but their materials more frequently mixed together, the isotopic ratios would simply be a uniform average. Thus, many large clouds may have a life of their own, forming stars which again expel gases back into the clouds and reform into stars over a long period of time without mixing very much with the remainder of the galactic material. This long cloud lifetime is quite surprising in view of the fact that they are dense enough for gravitational forces to pull them together into a collapse in times as short as about one million years. Of course, for a complete collapse to occur the clouds must be spherical so that the whole cloud collapses into one spot, and they are generally not spherical at all. Evidently, only small regions collapse at any one time, generating stars which then heat up and expand the clouds again. Details of why the cloud lifetime is so long are not yet clear. However, it seems from isotopic abundance measurements that these clouds must be very old, maintaining their compositional integrity over a long period of time. Furthermore, if they did collapse in any very much shorter time, we now know that the clouds are so abundant that they would have formed an enormously larger number of new stars than are actually found in our Galaxy.

In contrast to the $^{13}C/^{12}C$ ratio, the $^{18}O/^{17}O$ abundance ratio is remarkably uniform throughout our Galaxy, as shown in Table 5. Here, even in the clouds which are quite close to the galactic center where the ^{13}C abundance has been very much enhanced by fairly rapid production of stars and nuclear energy, the ^{18}O/^{17}O ratio is quite similar to that in less active regions. This is not so surprising in view of the fact that both ^{17}O and ^{18}O are rare isotopes and thus, if they are both formed from more abundant isotopes in essentially the same process throughout galactic history, their ratio should stay approximately constant. The surprising part of Table 5 is that the abundance ratio for these rare isotopes is quite different on Earth from what it is in all the other clouds so far examined. In contrast to the ^{13}C data, this would indicate that material from which the Earth was formed had some rather special and peculiar history.

One other isotopic abundance ratio is noteworthy. This is the abundance of heavy hydrogen, or deuterium compared to ordinary hydrogen. The spectra of dark clouds show that the ratio of deuterium to hydrogen abundance decreases towards the center of the Galaxy. This gives some substantial indication that stellar activity tends to destroy deuterium rather than produce it, information which seems to confirm the idea that deuterium was formed in the initial big bang of the universe rather than by nuclear reactions in stars. Thus the deuterium we see is a remnant from that big bang, and its abundance presumably gives us sound information about that very beginning of our universe, which also supports the remarkable conclusion that our universe may continue to expand indefinitely.

While the temperature established by molecules for various clouds has already been discussed, it needs to be emphasized that generally this is the temperature of kinetic motion of individual molecules within the clouds, and by no means represents an overall temperature equilibrium. The population of excited states of molecules, for example, is commonly far from a temperature equilibrium value. Molecules are in contact with various heating and cooling sources. Collisions of the molecules observed with hydrogen represent one source of heat. Infrared radiation from stars within the cloud and from the dust warmed by stars represents other sources. Cooling of molecules is most commonly provided by radiation of the molecules themselves into space. It is a well-known principle of thermodynamics that if a machine or system interacts with two different heat sources, one warm and one cooler, work can be produced which can either heat the system hotter than either source or refrigerate it below the temperature of either one. Such natural heat engines are common in interstellar space. For example, the 6 cm resonance radiation of formaldehyde shows that formaldehyde excitation has been cooled to a temperature as low as one degree absolute, even though everything else with which it interacts is definitely warmer. Still more spectacular is the very intense and hot radiation from natural masers occurring in interstellar clouds. As is known from laboratory work with masers, if an excess population of molecules occurs in an upper energy state compared with those in a lower energy state, the molecule may give up its excess energy to an electromagnetic wave, amplifying it. If the amplification is very high, it can produce radiation of such intensity that it represents an exceedingly high temperature. Certain molecules in interstellar space and in the regions immediately around stars or in nearby clouds commonly produce very powerful amplifiers. These include water, methyl alcohol, the free-radical OH, and silicon oxide. Figure 10 illustrates such radiation, the H_2O spectrum of the dark cloud and ionized region known as W49. Each individual peak represents amplification by water of radiation at a particular frequency, corresponding to the Doppler velocity of the particular cloud of water vapor which provides the amplification. W49 is quite distant from us, almost at the other side of the Galaxy. Intensity of the radio waves we receive from it is still relatively high and can easily be detected by rather simple antennas and receivers. Since W49 transmits at a single molecular frequency energies which are comparable with the total radiated energy of the sun—about 10^{26} watts, it could be described as one of the most powerful radio broadcast stations known. The region emitting in this case is large, but not as large as that of one of the big molecular clouds. We know this since the amplitude of the emitted waves varies during times as short as about one month. This can happen only if the size of the object is not much larger than the distance radio waves travel in a month—quite large but still smaller than the larger molecular clouds. Such microwave amplification by water is powerful enough that we can locate maser amplifiers even in other galaxies millions of light years away.

It is curious to consider that naturally occurring masers have been operating in space for billions of years, and are relatively easy to detect, but were found only after the invention of masers on Earth. They were invented by nature much earlier, and if man's study of astronomy had been more assiduous, he would have thereby discovered masers and lasers sooner—what a strikingly practical discovery might have come from examining the heavens!

Other very intense energy generation is manifest in shock-waves within the large molecular clouds. Shock waves on an astronomical scale have been studied theoretically for some time, but now we have direct evidence of their existence and measurement of some of their properties. Such waves may be caused by stellar explosions within a cloud, by pressure resulting from intense heating of part of the cloud by ultraviolet radiation from newborn stars, or simply from a partial gravitational contrac-

MICROWAVE WATER VAPOR EMISSION

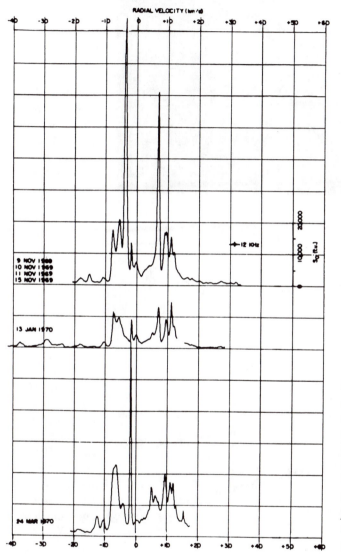

Figure 10. Radiation produced by water maser amplifiers in a region known as W49. Variations in intensity over a period of a few months are very noticeable.

tion of the cloud which accelerates part of it with respect to the remainder, producing violent cloud collisions and shock waves. Evidence for large scale shock waves in the giant Orion cloud have been found from detection of the spectrum of very hot hydrogen. This large dark cloud lies just behind Orion's well-known bright nebula shown in Figure 11, its center being indicted by the arrows. Within this dark cloud there is even more in-

Figure 11. The bright nebula of Orion. Behind it is a dark cloud, whose center is marked by the arrow.

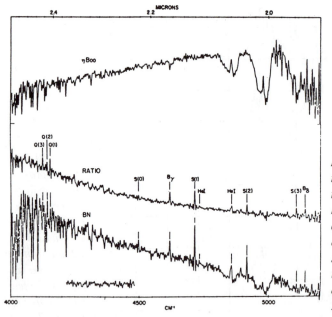

Figure 12. The rotation-vibration spectrum of hydrogen molecules in Orion, emitted from a shock wave temperature approximately 2000°K.

teresting activity than is seen in front of it. Figure 12 illustrates a spectrum of hydrogen from this region, showing the 2 micron radiation which corresponds to rotational-vibrational transitions of the hydrogen molecule. To excite this radiation, temperatures of about 2000° are required. Yet we know from other molecular evidence that the bulk of the Orion cloud is not warmer than about 70° absolute. Maps of this 2 micron radiation in the Orion cloud give a picture of the size and distribution of the shock

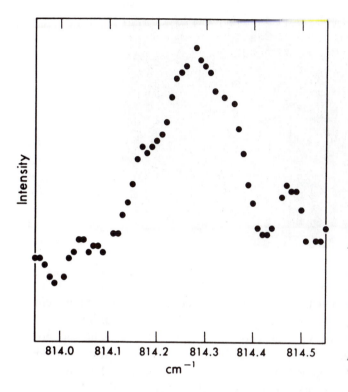

Figure 13. A pure rotational line of hydrogen molecules, emitted from part of the shocked region in Orion which is at temperature about 600°K.

wave, which turns out to be quite extensive though not very thick. Behind the shock front, the gas cools somewhat, leaving in its wake a large amount of warm gas. Figure 13 shows a line of the pure rotational spectrum of hydrogen from this part of the shocked region. This line appears strongly in regions where the temperature is about 600°. It is remarkable that radiation from hydrogen molecules can be detected in view of the fact that the hydrogen molecule has no dipole moment and hence can radiate only very slowly as a result of its minute electrical quadrupole moment. Thus, while H_2 is the most abundant of all molecules in interstellar clouds, the rotational spectrum of Figure 13 represents the first detection of this radiation from hydrogen outside the solar system. The precise origin of the shock wave discovered in the Orion cloud, its motions and the high temperature chemistry within it are still something of a puzzle, and will gradually be unraveled as additional molecular transitions are detected and studied.

These many new discoveries have come about in part because of an increased awareness of what is present in dark clouds and of their importance. In part, however, they have been made possible by the development of new technology and instrumentation—an illustration of the very close

Figure 14. Photograph of the Very Large Array (VLA) of antennas in New Mexico which is coming into use for very high angular resolution.

coupling between basic science, engineering and technological development, and the production of sophisticated components and equipment by
industry. Future extension of this work and further revelations of what is
going on will in my view depend very much on future technical and experimental developments. Fortunately, we can be sure that the immediate
future will be enlightened by additional equipment and techniques just becoming available. Some of these will be described.

Figure 14, for example, illustrates the Very Large Array (VLA) of radio
telescopes which is on a high plane in New Mexico. This instrument is
exceedingly sensitive because of the number of large antennas. Used in
unison as an interferometer these antennas allow detailed spatial study of
molecular distributions in clouds or around stars, thus giving us a more
precise picture of the distribution, turbulence, and detailed structure of
molecular clouds.

Infrared heterodyne spectroscopy is a developing technique which provides an opportunity for very high resolution of molecular behavior both
spatially and spectrally. I shall emphasize here high spectral resolution because easily visualized progress in this direction has already been made.
Infrared heterodyne detection is essentially the same in principle as the
heterodyne detection and amplification common in radio receivers. It involves a local oscillator near the frequency of an electromagnetic wave
which is to be detected. The local oscillator wave is mixed with a signal
and then the difference frequency is further amplified at lower frequency.
Such work is now possible in the infrared as a result of infrared lasers
which can serve as local oscillators and the development of special semiconductor materials which can respond sensitively and rapidly to infrared

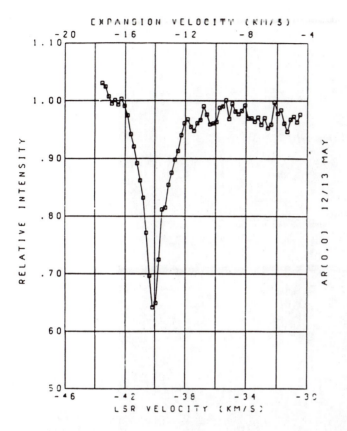

Figure 15. A spectrum of ammonia showing several closely-spaced velocity components. The spectrum is in the 10 micrometer wavelength region, and obtained by infrared heterodyne detection.

radiation. A combination of these two has allowed, for example, the detection of ammonia expanding in a small region around the infrared star, IRC+10216. Figure 15 illustrates a spectrum obtained with this new technique, showing absorption due to ammonia. The several valleys in this spectrum are due to ammonia gas at several velocities differing by only one or two kilometers per second in velocity as the material expands around the star. This gas can also be traced far away from the star as it expands to provide molecules for and to join some interstellar cloud. While this particular work was done in the 10 micron region, infrared heterodyne techniques should be all the more valuable as they are developed for the somewhat longer wavelength infrared.

Figure 16 illustrates NASA's Kuiper Observatory in a large C141 airplane for work at about 45,000 ft. above most of the Earth's atmosphere. This plane contains a telescope of about one meter diameter which looks out the side of the plane and points in a stable direction as the plane maneuvers. Its primary function is to fly high above most of the water vapor

*Figure 16.
NASA's Kuiper
Observatory, a
C141 airplane
which provides
a high-flying ob-
servatory for
measuring radia-
tion absorbed by
the atmosphere
before it reaches
the Earth's sur-
face.*

in the atmosphere because water absorbs much of the infrared impinging
on us from outer space. It thus allows astronomical work in infrared regions which are blanked out from our view on the Earth's surface. New
instruments are steadily being developed for this high-flying observatory,
both for the far-infrared region (50–500 micrometers wavelength) where

Figure 17. An infrared observa-, tory being considered for use in the Shuttle, providing a telescope in orbit for detecting radiation strongly absorbed by the Earth's atmosphere.

molecular rotational spectra should be evident and for the shorter infrared regions normally absorbed by water vapor.

While the airborne observatory can give us better visibility of external space, at certain wavelengths even the thin atmosphere above 50,000 feet screens out radiation—in particular it interferes with ultraviolet radiation and certain types of very sensitive infrared measurements. Hence there are already in operation telescopes in satellites for ultraviolet observations, the International Ultraviolet Explorer and the Copernicus satellite. These have yielded some very interesting information about molecules in the thin clouds where dust is not too plentiful. An infrared satellite is also far along in the construction stage. This satellite will provide valuable information, but only about the broadband heat radiation at various wavelengths. Further molecular studies require somewhat different instruments, considered for launching with the space shuttle, as illustrated in Figure 17.

With the extensive base of what has already been discovered, the very promising new instrumentation which it appears will become available, and with the many outstanding young scientists in these fields, we can look forward to very exciting future discoveries.

PROCESS OF
DISCOVERY

Science, Technology, and Invention: Their Progress and Interactions

T he interaction between pure and applied science is most frequently thought of as a flow of ideas and contributions from basic discoveries toward applied science and industry. As scientists we are both conscious and proud of such contributions. A simple documentation of this well-recognized type of flow is provided by the list of Nobel prizes in physics since 1925; approximately 40% of the discoveries involved led rather rapidly to substantial industrial applications of the new ideas they provided. Most of these discoveries can be said to have originated in basic science, and rather soon thereafter many have had important industrial applications; some have produced large industries. One can point, for example, to the discovery of the neutron by Chadwick, coupled with Cockcroft and Walton's artificial radioactivity and slightly later Lawrence's contributions to nuclear processes. These led to the growth of a nuclear industry that has many important applications in addition to the production of nuclear energy. There is of course the transistor and solid-state electronics. There is also quantum electronics and the somewhat smaller but already substantial industry of masers and lasers and their applications. There are other discoveries that may hatch industries somewhat later, such as the Josephson effect, which has wide potential applications but it has not yet produced an industry. In addition to discoveries that have rather specific applications, there are also abstract ideas such as the uncertainty principle, which may have no obvious and specific application to industry but yet pervades all our thinking and thus makes many indirect contributions to applied work.

While scientists are rather familiar with the flow of contributions from basic to applied work, and the phenomenon is recognized in generality by the broader public, I do not believe adequate recognition is given to the reverse flow—the contribution of technology to basic science. Convenient and sophisticated instrumentation is but one example. More broadly, such contributions encompass important scientific discoveries produced by applied research, the development of industrial and commercial products on which much basic research depends, and new technical possibilities that emerge from applied work in industry and in military and space programs.

Contributions of Applied to Basic Sciences

Harwit has surveyed the growth of astronomy and, while his views may be somewhat controversial, I believe they make an important point that is pertinent to understanding interactions between the sciences, basic and applied. He concludes that many of the most significant discoveries in astronomy have been made shortly after a new technique has been introduced, and generally by people from outside of the field. One can cite a number of cases of this type; for example, Jansky's discovery of radio astronomy in the early 1930s. Jansky was an electrical engineer at Bell Laboratories who constructed a system for the sensitive survey of noise in radio communications. In the course of this work, he discovered and identified the first radio waves from extraterrestrial sources. Other examples are the initiation of x-ray astronomy by a group at the Naval Research Laboratory, including Tousey and Friedman, by the use of newly available V2 rockets; the subsequent discovery of x-ray sources outside of our own solar system by Rossi and Giacconi; and the first detection of γ-ray bursts by the Los Alamos group using the Vela satellite, which had been put in orbit for military purposes. There is also the work of Hewish and Bell, who put into operation new radio equipment that could detect short pulses of radio waves and unexpectedly discovered pulsars. Then there is the discovery by Penzias and Wilson of the microwave background radiation as a result of careful examination of microwave noise with receiving systems of new sensitivity. Many recent astronomical discoveries have grown out of the space program, a program stemming largely from the efforts and interest of engineers and government officials, but promoted to a certain extent by scientists.

Some astronomers appropriately say that, although these were important discoveries, the most important work of astronomy is the understanding of such matters and their coherent interpretation. Neverthe-less, in any list of discoveries that have fructified the science, these ex-amples are ones of prime importance and several have initiated new branches of astronomy.

None of the discoveries cited above was the direct or immediate result of the availability of some commercial product, but they do come from in-strumentation that largely depended on applied development and interests. The contributions of more or less standard commercial products are also exceedingly important to scientific discovery. Obvious modern examples are computers and solid-state electronics. Another is the development of a wide variety of commercial lasers, which are rapidly penetrating scientific laboratories and providing new experimental potentialities. Many types of specialized materials are very important to research. In my own present work I rely on infrared detectors developed largely for the military estab-lishment, on very fine wire mesh gauzes developed for commercial filter-ing, on sensitive solid-state amplifiers, and on very fast circuitry. These are all available with a quality that I could not practically produce myself. If they were not available, I would have to do a different kind of astro-physics and could move only at a much slower pace.

When Millikan undertook his famous oil drop experiment, he person-ally built about 1,000 lead cell batteries in order to have a high, stable voltage. We now get such a voltage supply off the shelf; it is simple, small, cheap, and quickly available.

Some of the contributions of industry come about because of a particu-lar industrial drive and interest; some are possible simply because of the availability of money. Industry can undertake developments and explora-tions that people in basic sciences and at universities simply cannot af-ford. This allows a very much broader base for developments that are diffi-cult, expensive, and yet important enough to industry that they are vigorously pursued. In the long run, they are often also important to basic science.

There are many interesting cases of the effect of adequate resources in the military or the space programs as well as those primarily associated with industrial interest. In the early days when integrated electronics were just being worked on, I remember a telling answer when I asked an ac-quaintance I met in Washington as to what he was doing there. He was ap-proaching officials in the Pentagon to get more money for his company to work on integrated electronics. He explained that integrated solid-state circuitry was so far off and chancy that his company would not think of putting any of its own money into such developments but, since the mili-tary was willing to support this type of research, the company was glad to be occupied with what would otherwise be a poor research investment.

Such a judgment was not true of all companies, but it was not an uncommon attitude. Without governmental interest and a massive infusion of money and effort, the integrated electronics that now means so much to both science and commerce might have been delayed.

I have emphasized the contributions from applied to basic science partly because of their importance, but partly because I believe they are often not adequately recognized. The flow of ideas from basic to applied science is not less significant, but is more frequently noted.

Importance of Interactions Between Sciences and of Concentration of Effort

In addition to the development of ideas and of technology, the progress of science is much affected by the connections and flow of information between fields and between scientists. One can see this in a positive sense from fortunate interactions that have stimulated ideas and in the negative from delays in important developments because of the lack of appropriate connections.

Consider, for example, the imaginative work of Gamow, Alpher, and Herman that suggested and examined isotropic radiation produced by an initial universal explosion. The impact of this work was limited, I believe, because of such misconnections. Gamov, Alpher, and Herman apparently did not appreciate the fact that we were very close to being able to detect the isotropic microwave radiation which they predicted theoretically; appropriate experimentalists who might have been interested, on the other hand, were not close enough to this theoretical work to become interested. It hence remained for Dicke to later reinvent the whole thing; he fortunately was knowledgeable about both the theoretical and experimental aspects. While Dicke was developing equipment to look for this radiation, it was found by Penzias and Wilson—one might say by accident, but of course an "accidental" great discovery depends on necessary scientific sensitivity and ability. Initially, Penzias and Wilson were not in touch with Dicke and his already-developed ideas. Their understanding of this discovery might have been slower than it was if they had not very quickly made such contact.

Another illustration of the importance of contacts comes from some of my own work. I apologize for so many personal illustrations, but of course I am a bit more sure as to what happened in my own field than I am in others. While musing on ways to make real my general idea for a maser, I was profoundly helped by a conversation with Paul, a German scientist who had visited Columbia University and talked about an interesting new way of producing molecular beams of especially high intensity. His method was fresh in my mind, which made it seem practical and possible for a maser to work. Without this conversation with Paul, I might possibly have dismissed the idea for some time.

Obviously, many factors enter into scientific or technical productivity. An interesting aspect of this comes from Shockley's study some time ago of the distribution of numbers of patents per staff member at Bell Laboratories and of numbers of scientific papers published per staff member. Of course, simple numbers of patents or papers do not necessarily measure creativity. However, they represent some measure related to productivity, and that measure shows a striking distribution. A few individuals in this study had each produced 100 or more patents, while most staff members had 1, 2, or only a few. The distribution was very peaked, as was the distribution of numbers of scientific papers. It clearly did not correspond to any simple gaussian function of, let us say, IQ, but was characterized by an almost divergent peak, approximated by what is generally known as a $1/f$ distribution. That is, the number of individuals who had produced a given number of patents or papers was approximately inversely proportional to the number of such patents or papers.

One can speculate about what makes certain individuals in their particular situations so highly productive; a statistical discussion by Montroll makes it very plausible that the combination of a large number of different effects, each making its own gaussian-type contribution, characteristically produces such a $1/f$ distribution. It also seems reasonable that many different individual factors affecting research productivity can have a multiplicative effect. One of these probably is intensity of two different kinds—the intensity of work on the part of the individual and the intensity of concentration within an institution, so that the cross-fertilization of information and ideas peak both in the individual and at the institution. An individual's intuition is clearly related to his or her knowledge, experience, and rapidity of sorting out better solutions from poorer ones, using these qualities. Frequently, the efficiency of sorting out unfruitful routes for solution of a problem depends on easy access to a friend who has relevant knowledge. Hence, being closely surrounded by other scientific and technical activity is often important to new and exploratory work. When an interesting but nonstandard idea comes up, the quality of access to apparatus for quick experimentation or the quality of opportunities for discussion with others who have useful knowledge or experience can make its exploration either practical or impractical.

We need to consider the extent to which we in the United States will lose the ready contact with others on the forefront of their fields or the ready access to information and materials needed, as research and industry of high quality continue their spread to additional countries and the United States plays a less dominant role in science as a whole. We have in the past been fortunate in having a remarkably central role in science and technology, in attracting to the United States many outstanding scientists from abroad, and in maintaining active industries in many fields. The dilution of the world's scientific and technical efforts can bring inefficiencies as well as the welcomed increased participation of all nations in the scientific enterprise. Fortunately, today communication is rather efficient and travel is quick and relatively cheap. However, travels to many meetings can become oppressive and it is not clear that intercommunications among scientists, particularly scientists in different types of environments, will in the future be as favorable for anyone as they have been in the past for the U.S. scientific community.

While the general environment has a profound effect on the development of science and technology, one must not neglect the importance of individuals and of the microenvironment. This bears on the question sometimes asked whether scientific or technical innovation is really the product of some uniquely innovative individual in the right situation or instead is simply a series of discoveries whose time has come. The latter represents the view that, when the time is ripe for a particular development, it comes along; while there is an individual or a group which produces it first, there are many others who might have done so only a short while later. This view would make discovery not critically dependent on any individual or fortunate environment but rather a matter of general overall progress. There are certainly some ideas and scientific developments with this characteristic, but one can also find those that were delayed a long time for lack of some key element or individual.

Discovery of the isotropic radiation (the "big bang") is one of those that was probably delayed for want of the right discoverer. Available techniques and ideas would have allowed its occurrence much earlier if someone had made the appropriate approach. The breakdown of parity is another case; effects of nonconservation of parity were noticed as early as about 1930, and someone with the right idea might have suggested such an effect much earlier, even if clearcut experimental proof would have been more difficult then than in the post-World War II period. The transistor, on the other hand, is a case in which effective discovery and development could probably not have occurred much earlier than when it did. There was at least one overt attempt before World War II to make a solid-state amplifier but it failed for reasons that were poorly understood at the time. Success depended, I believe, on the development of solid-state theory and more understanding of the solid state than was available at a substantially earlier time.

Antibiotics provide another interesting case. Extensive use of antibiotics came along only 10–20 years after their discovery by Fleming; nevertheless, there was a dead period of about 10 years with little interest in antibiotics until Florey began, in 1938, to look into their potential. Within only a couple of years after this, antibiotics became available for medical purposes. Thus, important gaps in scientific developments do occur and progress in science and technology is not a regular and steady march forward, produced automatically by the world's or a nation's generally favorable regard.

The Problem of Planning and Foresight in Research

Two different routes toward scientific progress may be described; they cannot be completely separated, but for heuristic purposes their separate consideration is useful. One route is the pursuit of clearly important science that is foreseen by a number of those who are expert in the field. In many cases, the importance of a certain line of development seems clear. Furthermore, it may be known approximately how to pursue the appropriate line of investigation. What is then needed is only the appropriate combination of money, time, and people. Those in the field recognize that the investigation is important, working at it with adequate intensity and ingenuity will provide answers to important questions, and the research will be great fun for the scientist. This is a very significant way of making new scientific discoveries, and there is much of science that fits such a mold.

Another route to discovery involves work not recognizable to everyone as important and perhaps even regarded as rather unimportant or infeasible by everyone except a few individuals. Even the few scientists who think work in the field is interesting may not have the right idea or approach initially. Nevertheless, the number of important surprises or discoveries from such situations is significant. Developments of this type do not necessarily involve more innovation than the first, more easily understood route. However, they do come on us more suddenly and are more surprising, and we need to appreciate and be sensitive to their importance. The more obviously important lines of development for science may be carried out with a high level of imagination and skill. Nevertheless, it is generally not such a surprise if they achieve important results. The unexpected development from an obscure or apparently mundane field and the obviously important line of investigation are simply different routes to discovery; the latter are much easier to talk and reason about but to overlook the "unexpected" route would be wasteful and stultifying.

Particularly in discoveries whose very occurrence is difficult to foresee, an interesting and valuable interplay between pure and applied science is likely. This is partly because the unexpected, happening in either pure or applied work, may contribute most importantly to some field other than its own. Discovery of radioastronomy or of the big bang radiation in the course of applied work are such cases. I hope I can appropriately refer again to the development of quantum electronics, one of my own interests. The development of a very intense, coherent, and exquisitely controllable light beam depended, surprisingly, on intensive research on the interaction between microwaves and molecules. This is because the laser grew out of the maser, being a particular kind of maser adapted to IR and optical wavelengths. The maser itself came from microwave spectroscopy. Microwave spectroscopy was in turn a somewhat surprising development that came out of interplay between pure and applied science. It began largely in our industrial laboratories because that was the locale of microwave technology. However, the industrial laboratories were not terribly interested and hence such research soon moved to the universities where it proceeded for some years before leading to invention of the maser and laser. Particularly after invention of the laser, industry moved vigorously back into the field of spectroscopy and quantum electronics and has done a remarkable job in subsequent developments. However, the dozen or so earliest contributors to the field were almost entirely scientists who had grown up in the field of radio and microwave spectroscopy. At a later stage it was developed rapidly and imaginatively in industry by somewhat different personnel including, of course, many engineers. This development is in turn now supplying high-quality instruments that are invaluable to basic scientific research.

While individual scientists may at times be remarkably perspicacious in evaluating a particular direction of scientific development, more generally the problem of foresight in research and technological development needs to be approached with a clear recognition of the limits of our wisdom and with rare modesty. As support for this view, I want to comment on two studies of the past that were directed toward foreseeing our technological future. One was produced by a high-level committee appointed by Roosevelt, which made its report in 1937 on technological trends and national policy. In this report, the committee looked back at other predictions, commenting in particular on those made in 1920 in the *Scientific*

American, which undertook to predict the coming 75 years. The commit tee's view was that the *Scientific American's* predictions were fairly good. Fifty percent of its predictions were said to have been correct, 25% were still doubtful, and only 25% were wrong. However, if one examines what was actually predicted, I do not think the results are especially impressive. On examining this prediction, I noted in the *Scientific American* of the 1920s an article entitled "How science has now made gambling on petro-leum prospects a thing of the past." Now to the predictions.

The specific predictions of 1920 made some very reasonable com-ments—for example, in the field of genetics it said, "We shall continue to pick up information as to how heredity works.—Doubtless the goal here is the human animal; whether that goal will be obtained is a matter of guesswork." But it also said, on the subject of extrasensory perception, "No careful person can categorically deny the accumulating evidence that there is really some sort of communication between individuals widely separated in space." With this kind of generality and batting average, I am not highly impressed with a count of 50% correctness. The 50% correct predictions tended to be things that were already under active considera-tion; it's what was missed that is impressive. The misses included any mention, for example, of radio broadcasts (which came on stream with the opening broadcasts of KDKA only one month later) or of talking pictures. However, predictions of the Roosevelt-appointed committee are more to the point because they were made by experienced scientists and engineers of the type the National Academy of Sciences might assemble for this purpose.

The report to President Roosevelt in 1937 was intended to allow sound thinking about the future impact of science on society and was done by appropriate and well-known scientists and technologists of the time. The Academy had itself appointed another committee that was interacting with the committee Roosevelt had appointed especially for this purpose, so that a still wider group of scientific statesmen was represented. The re-sulting report made a number of wise assessments: it said, for example, reasonable things about agricultural research and the development of agri-culture and about the development of better rotating machinery, which would make creation of electricity more efficient. It also discussed syn-thetic gasoline from coal and synthetic rubber as having substantial poten-tial. However, again what the report missed is more interesting. It missed almost all of the most exciting developments of the next few decades. It missed nuclear energy, though this possibility became obvious only about 1 year after the report, with the work of Hahn and Meitner. There was no mention of the field at all but, if the committee did think about it, mem-bers might have referred to a statement a few years earlier by Lord

Rutherford, who was clearly the expert of the time. Rutherford said "Anyone who expects a source of power from the transformation of these atoms is talking moonshine." In the newspaper report of the discussion following Rutherford's statement, it was essentially only E.O. Lawrence who was brash enough to say that, while he did not know how it could be done, there still might be a chance. The committee missed antibiotics, which had been discovered by Fleming some 8 years before that; Florey's work was only 1 year in the future. It missed jet aircraft. It missed rocketry, space exploration, and any use of space. It missed radar. It missed computer development. It missed the transistor. It missed quantum electronics and, looking still a little further in the future, it missed genetic engineering.

If one lists the most interesting and exciting developments in technology during the next decade or decades, the ideas that the committee missed approximate such a list.

The above examples should give us all pause as we write our committee reports. Having just finished participating in one on the future of astronomy, I must share concern over the probable extent of our foresight. The problems with scientific decisions made by committees are, I believe, quite deep. Peer review seems to me to have serious problems with respect to innovation. Committees tend to be conservative as a group and to follow established routes. These may be good routes and result in good science. However, for many of the developments that bring important surprises one needs the spark of occasionally crazy enthusiasm that comes from an individual. Someone needs to have the conviction that a new approach is right when others, perhaps even the community as a whole, are skeptical. Wise committee members, particularly senior statesmen, all too easily recall "Yes, I thought of that 15 years ago; I looked at it and it's not going to work," without recognizing that there are new aspects to the idea that may allow its success.

The obvious and difficult problem of supporting offbeat ideas is how to sort out the real nuts from the apparent nuts, because there are plenty of people who are enthusiastic about something with which no one else agrees. To this problem there is no straightforward nor completely reliable solution. We must search for ways in our national planning to see that there is enough looseness to allow both error and support of promising but uncertain research. Some of the necessary looseness is, unfortunately, much dependent on the supply of money and, at present, that is especially difficult. As the supply of money becomes tighter, committees and decision makers tend to become more conservative, feeling obligated to put the limited supply of money on those things that are most clearly going to pay off. They also know that money for their own personal work is in short supply, which produces an almost unavoidable tendency toward conservatism. When financial support is more abundant, there is more taste for directing some money toward the longer research chances or less obviously rewarding approaches. This does not mean, however, that such a policy is necessarily less shortsighted when money is tight than when it is plentiful.

By way of summary, I have a short list of some social requisites for scientific and technical innovation.

First on my list is a general interest throughout society in intellectual ideas and a sense of the excitement and value of discovery.

Second is a diversity of approach, a diversity of types of institutions, and of people. This means support of research, types of research, and individuals that are not obviously in the mainstream of important science.

Third, along with diversity and widespread diffusion of effort, there must be at last a modest number of institutions where the research environment is very intense. That is, there must also be concentrations of excellence, implying concentrations of skillful people who maintain high standards and interact in a supportive and inspiring way.

We need the diversity that comes from research efforts in a wide variety of institutions with various traditions and approaches. However, we cannot rely on diversity alone. There must also be institutional peaks that are particularly creative and may well appear to dominate the scene. To combine this with diversity requires a particular overt openness with respect to funding so the less well-recognized and the offbeat work can be supported. This may require overlap of the charters of several sources of funding. As an example, I am skeptical that the nation would be well served if the National Aeronautics and Space Administration and the National Science Foundation were to map out too carefully the type of astronomy each should exclusively support rather than allowing substantial overlap. If only one organization is responsible for making decisions in a given field, over a long period of time there can easily be a stagnation of ideas and approaches. More than one kind of authority to which researchers can appeal when turned down improves the chances of breaking down overly rigid decisions.

My fourth condition is the encouragement of interaction between sciences and, in this interaction, I would emphasize the interaction between pure and applied scientists. Easy and common contacts between individuals in universities and between scientists in universities, industry, or government establishments is an important phenomenon that needs our attention. Unfortunately, during the recent past our universities, industry, and governmental establishments became badly separated. They are now slowly becoming reconnected. This brings some difficulties and conflicts of goals but also great values. One of the more obvious present examples of this important aspect of our scientific and technological success is in bioengineering, where both the value and the uncertainties in connections between academia and industry are obvious. One can regret the distraction that such interactions with the larger society brings to universities. One can regret the administrative complications. One can perhaps even regret that there are concomitant opportunities for personal wealth that may stand in the way of normal scientific values and procedures. Nevertheless, such new drives and new connections will no doubt be very fruitful for innovation and, in the long run, the nation and its science will benefit from maintaining and encouraging such interactions between the nation's wealth of scientists, pure and applied.

The Possibilities of Expanding Technology

Progress does not seem to be slowing down and may instead be accelerating. This raises the question: how far do we have to go—how far will knowledge be taking us? Is the realm of technical knowledge infinite, or will our job of exploring science and technology sometime be complete? One might, for example, imagine that knowledge is like an island which we approach from its shores. We explore all the island and then know what is there: the physical universe might possibly be like that. Or we might imagine expanding knowledge from a center, so that as we push back the boundaries, exploring continually, the boundaries simply enlarge. As we progress we can see a little bit further into such a universe, but there is always more in front of us than when we started, even though we have learned much more and explored a great deal of territory.

My view is that knowledge and technology are more like the latter case. Consider, for example, the world of physics. When I went to school the most fundamental and, we thought, simplest part of physics, the fundamental particles, was comprised of only a few elements: the electron, the proton, and then the neutron. Further examination showed that these are made up of further systems, which in turn are made up of further ones and, as far as we really know, the particle physicists simply proceed into an unknown region of apparently growing size. Even in this relatively simple and fundamental area, no limit has yet appeared and our progress can fortunately continue. This has been an exciting and exhilarating century, and one can hope that there are not limits to the contributions scientists and engineers may make in the still longer run to the knowledge and scope of the human race, and to its welfare.

My second point it that our field is a very young, bumptious, and still maturing one. If, for example, we say that the modern aspects of electrical engineering were born 100 years ago, I suppose we are now only in late adolescence or perhaps young maturity. The field is growing, becoming stronger, and is ready to take on still more responsibilities. With the present characteristics of the field and its importance to our civilization, those increased responsibilities seem destined. If we are in adolescence after the first century, there are still many centuries ahead. Any nearby limitation will not, I suspect, be due to the inherent nature of science and technology: the limitation may be us. We may just develop into good sedate members of society who allow our field to advance very successfully into a middle-age slump. That clearly has not yet happened. Nevertheless, such age effects can occur and we must be alert to them. How can they occur? Consider that the United States has been blessed with rapid development, even an explosion of developments, which have made us feel optimistic about the future, free to speculate hopefully, ready to experiment, to accept new ideas, and to welcome people from all fields into our own or to transmit ideas rather freely across all boundaries. That is an important part of what has made for the greatness of American science and technology. It has also required a willingness to take chances.

In the business world, I think everyone recognizes the importance of the small entrepreneur who takes chances and the entrepreneurial company which, while it may fail completely, will on occasion succeed beyond the wildest dreams. We also recognize in principle the importance of letting everybody try new ideas and of supporting research which explores the unknown. However, it is much harder to be convincing about the latter. The reason is that research takes so long to develop into visible uses. I would say, for example, that the transistor really began in the 1930s when Mervin Kelly, at Bell Laboratories, decided to support substantial fundamental work on the new solid-state physics. The actual invention came later, and any real applications success at least 20 years later. This is common with research results, which cannot be completely predicted. And it is the surprises and the unpredictable that often produce the real breakthroughs. But the unpredictable is very unsatisfactory for the politician or even the corporate officer to support. We all like to try to figure out the consequences of our decisions. We also like to see results happen within our term of office. In the field of research, where one cannot know, this puts us in a difficult position. Hence long-term research, the wild chances, the freedom for individuals to try ambitious but uncertain ideas, are always difficult to support. I hope we can avoid the middle-age slump of overcaution, or of planning only for the plannable, and support a reasonable fraction of the long chances which may develop only slowly if at all, the small and off-beat operators, or the new ideas that not everybody believes in. This is what will ensure that the spirit of youth and growth can continue.

The third point I would like to make is that the future is very much affected not only by the nature of our field, but also by human aspirations. Almost anything can be done that is not contrary to some basic physical law. What actually gets done is primarily determined by human aspirations. In that sense it is often interesting and revealing to look at science fiction, which in a certain sense represents humanity's dreams. The science fiction of the Greeks about the flight of Daedalus took a while to come true, but is now eminently real. *20,000 Leagues Under the Sea* came true a little more rapidly. Mankind wanted to go to the moon, and we did. Those things to which we really aspire, even if apparently improbable, are actually likely to happen. I sometimes think that the interest in laser beams is associated with Zeus' bolt of lightning and Buck Rogers' ray gun. Those were some of mankind's dreams, too.

In our aspirations, however, there are frequently conflicts. There are the aspirations of individuals or of nations to be completely free and independent, to be able to make their own decisions, but also the aspiration for safety, fairness, and orderliness, which all require organization of society. More starkly, we must recognize that there are even aspirations for dominance and control of other people in parallel with that for freedom. Such mixed and conflicting aspirations give us deep trouble and uncertainties, and are partly responsible for the terrible problem we have with the threats of war. The third world also has its critical aspirational conflicts. There is the desire for a higher standard of living, with abundant natural resources for each individual, but at the same time aspirations for procreation and many offspring. This dilemma of the third world and its poverty are only enhanced and emphasized by the richness of technology. Technology, training, and education are themselves a form of wealth. We shall become, I am convinced, less and less dependent on the normal products of the third world as technology makes mankind more flexible in the use of alternate resources and less dependent on what goods and services the third world has to offer.

How can talents be unleashed and made fruitful in further manifestations of the type of creative surge we technologists have experienced during the last century? Certainly we need to think carefully about our aspirations and where they lead. With appropriate and self-consistent aspirations, I believe that the human being's increasing understanding of the physical world and the potentials of technology make almost anything realizable during our next century.

The Role of Science in
Modern Education

I f I told you that science itself is important, you wouldn't be surprised. You might be a little more surprised to know that somewhat more than 90 per cent of all scientists of history are living and working to-day, and that the size of the scientific effort doubles about every ten years. If we make an extrapolation based on these past figures (sometimes a very unscientific process), we find that, by the time some of our students here become the age of some of us on this platform, every man, woman, and child in the United States will be engaged in a scientific pursuit. Now that may seem silly, and perhaps it is. But I'm convinced it contains an important element of truth. Science for the last hundred years has been the most revolutionary force in our society and probably will continue so for the next hundred years. It stands now at something of a crossroads. Science has grown to a size and complexity which are difficult for us to han-dle. It has grown to an importance in our daily lives which demands the attention of all citizens. Each member of society must somehow become a participant in the scientific enterprise, not just a bystander or beneficiary, but a participant and judge of how we should proceed and use our scien-tific knowledge. For this, some understanding of science on the part of every college student is tremendously important.

Science itself is a little hard to define. The nature of science might be described, I suppose, as a systematic use of intelligence in order to under-stand the universe around us, including man himself and his place in the universe. It is not a neat package. The so-called scientific method is in fact a bit of travesty on what scientists really do; I prefer Professor Bridgeman's definition of the scientific method. As a physicist and phi-losopher, he commented that the scientific method is to work like the devil to find out the answer, with no holds barred—and that is about how neat this intellectual effort is. But it is a response to man's curiosity, his sense of wonder, his drive to understand his universe and himself.

Science is not simply an attempt to answer esoteric questions in a strange laboratory off in the distance. It, in fact, is all around us and demands our awareness. There is a world of science in a blade of grass, most of which we do not understand. Look at this beautiful sky and let childish wonder ask, "What make the sky blue?"—a very simple, homey question, and yet one which leads us on almost indefinitely. The sky is blue because the blue rays of the sunlight are scattered by molecules of the air more than the red rays. How many of you have looked at that blue sky with polaroid glasses and found that the light is polarized? It is! And because this scattered light is polarized, bees find their way to honey when the sun is overcast—they wear polarized eyes. The same scattering phenomenon makes our sunset red. The red rays get through the atmosphere better than the blue rays. This is also the reason why radio waves can travel through the rain and clouds through which we can't see. Astronomers find that in certain directions the stars themselves are reddened by a similar effect, showing us that there is dust—stardust of some kind whose nature we don't know—between us and the stars, which tells us something more about interstellar space. And so man builds idea on idea and discovery on discovery.

The molecular theory was initiated, at least in some vague way, by the Greeks and Romans. In the last century we began to understand gases and their molecular properties a great deal better through the work of the Scottish scientist Maxwell and others. There were also enormously important English contributions to the chain. Newton had studied the different colors of light and Thomas Young sometime later helped us measure the wavelengths and understand light still better. It was still later that Rayleigh explained this light scattering in terms of molecular and gaseous properties, and the wavelength of light. The Austrian, Von Frisch, studied the dance of bees at their hives, and learned their dependence on polarization of scattered light. The Indian physicist, Wickramasinghe, has been working hard on understanding scattering of light by interstellar dust and has made progress. But students and professors are still puzzling over this dust and what the colors they produce by scattering can mean.

So you see the accumulated effect of science and how much it can tell us. Aristotle has commended that "The search for truth is one way hard, and in another easy, for it is evident that no one can master it entirely, nor miss it wholly. But each of us adds a little to our knowledge of nature and from all the facts assembled there arises a certain grandeur." Along with grandeur, science also brings simplicity, by making complex things eventually simple for us. It is like learning how to swim. You might struggle with science courses and wonder just where they lead and whether you'll get there. Once understood, things become simpler to us. Once we know how to swim, it seems natural enough and can be enjoyed thereafter.

Science also unifies the different parts of our experience. You may have the impression, as many moderns do, that science is becoming more and more specialized and separates people into their own small corners. In fact, I believe it fairer to say almost the opposite. It is true that science itself is enormously diverse, and people in science are highly individualistic. Scientific activities vary. They are as different as the banding of seagulls and study of their travels is from the lonely, hard thinking of a mathematician with paper and pencil, or from the struggles of the high-energy physicist with an enormous, big machine, as he studies the tiniest bits of matter. Yet as man understands more, he finds that in fact science brings diverse subjects together. As we get closer to the fundamentals of nature we find that physics, chemistry, or geology all have the same bases. I believe there has been no time for a century or more when it has been easier or more exciting for physicists, chemists, biologists, and astronomers to talk together. Even the economists and social scientists are finding much in common with the natural scientists.

I have spoken about the homey aspects of science. It is all around us— we have only to look and let our curiosity lead us. But such curiosity and science also lead us to deep philosophical questions. Here too, I would like to ask a simple question about the sky. Why is the sky dark at night? Now that is not a catch question, but a serious one. You may answer that obviously the stars aren't very bright, there are not many of them and hence if the moon isn't out, it is very dark, but let us think a little bit more about the stars and their numbers. If we could see indefinitely deeply into our heavens, and if the stars extend indefinitely, we would see more and more of them. We should see them just as we see a snow storm—individual flakes close by but, as we look in the distance, we only see snow as far as we look. In the same way, if we look far enough, why should we not see only stars and the whole heavens be bright? That the sky is dark is in fact a remarkable puzzle. It is explained, we think now, because the stars at some distance are receding. Their apparent characteristics are changed so that at some point we just can no longer see them, not because they are so far away, but because they are receding so fast. The result is essentially an edge or a limit to our universe. Thus we get readily from simple questions into cosmology and the nature of our universe, or further into deep questions bearing on man's nature, philosophy and his basic beliefs. Recently man has been able to detect a type of radio radiation in space, which seems to be a remnant of the past and thus take us back to the very birth of the universe—the first few days of an enormous explosion. This radiation is fossil radiation, if you like, which we can measure today and from it hope to understand the nature of these first few hours. There are still other keys to the enticing and important questions of the origin of our universe. If we turn from the largest scale now to the smallest conceivable objects, we come to the realm of high-energy physicists who are probing the ultimate particles of matter. As they examine the smallest particles, they find more and more pieces, and as they remove pieces they find that the pieces themselves are made of some of the original particles. So that perhaps everything is made of everything else. What a puzzle! The fundamental nature of matter and its structure may be already at our fingertips, or may escape us for many years. But we have confidence in at least steadily improving our understanding of these basic building blocks and the real nature of matter.

In the field of biology, there are also enticing and important questions. Some involve simply description and understanding of life. Others involve questions of the origin of life—how life came about and whether life has been created more than once, so that the earth and all its creatures are not unique. We are beginning to see possible ways of answering these fascinating questions. The function of the higher nervous system is another field which is steadily being explored. It is odd for the human brain to try to understand itself and one might wonder whether such a process is even possible. But at least we can understand parts of it. We still don't understand much of the complex machinery of the brain and yet a surprising amount of it is beginning to come into view. Ultimately perhaps we will understand how the elements of human personality are structured in the brain and whether values themselves are set by the interconnection of neurons.

The great sweep of science, the importance and fascination of the questions it approaches, and yet the simple steps with which it can initiate approaches to such questions make one wonder whether any individual can refrain from joining the researchers, or can afford to be left out of such adventure. Certainly anyone who considers himself liberally educated— liberally educated in the sense of having his mind freed, open to ideas, and sensitive to the world around him—should try to grasp some of these things and to enjoy the enlightenment which science can bring.

There is another side of science, equally important but in a material way—applied science, which has so much affected our daily lives. Basic science and applied science, the understanding of things and the application of this understanding, must go hand in hand. How they develop together can be easily illustrated, but not really understood fully. Attitudes of society toward science and its applications and their use have a profound effect upon civilization. Their successful development is not automatic, as one can easily see by looking at other civilizations which have failed in this respect. Think of the Egyptians. In some respects they were marvelous engineers, technicians, and even scientists. They understood a great deal of chemistry, their glass and ceramics are magnificent, their architecture is striking, they had civil engineers, and they did some mathematics. They saw and solved the practical problems of their day, and from a technical point of view they made a successful civilization—a civilization that lasted over three thousand years, a record none other has achieved. But Egypt stagnated and fell at least partly because, having solved civilizations's problems as they saw them, Egyptians did not press on with new knowledge and new ideas. They didn't have the sense of curiosity, nor the philosophical nature of the Greeks which eventually led toward development of Western Europe and the present blossoming of science. Look at the Mayans as another example. Mayan civilization was impressive and extraordinary. The Mayans knew mathematics and astronomy. But they failed after a few hundred years, and never caught the upswing of technology—they had not used such crucial technology as the wheel, or metal implements. Yet they knew the wheel—they used it in toys. They had metals as well, and used them as decorations. They somehow did not grasp their importance nor the ways in which one could apply some of the ideas which they had. In China, there was also a striking early civilization and a substantial development of science and technology. Yet Chinese civilization never reaped the benefits of a continually growing technology. The reasons are obscure, but may have been connected with organizational and manufacturing habits which were unsuited for such growth. Western Europe, the United States, the Soviet Union, and Japan all have met with success in modern times in the fostering of technological growth. Just what is required is easy to guess at, but hard to be sure of. Certainly the attitudes and values of society have a great deal to do with this continuous forward thrust.

To show the meaning of technological development in the most elementary, if also the most material way, consider its purely monetary effects. Much of our eight hundred billion national product is associated with the development of science and technology. For the United States to maintain a growing standard of living, or even to maintain a standard of living appreciably better than those in some of the less fortunate countries, it must be based on skill and education, including in particular good science and technology, and good industrial management. Our competi-

tive position as well as any possibility we may have of sometime bringing
the rest of the world up to the kind of standard of living which we would
like to see is much dependent on our abilities in these directions. And
such possibilities certainly exist. Consider nuclear energy, which is steadily becoming cheaper and will last us almost indefinitely. We know some
of the discoveries in medicine will have increased human longevity and
allowed us to do striking things in curing disease. We know some of the
improvements in agriculture which have steadily improved the output of
our agricultural workers. We know the changes in communications and
travel which have shrunk the size of our world and forced us into a close
neighborly relation with all other nations. We have before us a real ability
to remake the world almost in any form. But we must also have the understanding, the appropriate sense of values, the vision, wisdom, and the
willingness to work for it. We can make plans not only for the world, but
the solar system as well. One can think of lunar colonization, perhaps
even of producing an atmosphere around the moon, or a suitable one
around Mars, if we find that useful and desirable.

We also have the ability to remake man, an indication of one of the
most frightening and yet most important developments—the revolution in
biology. Biology is undergoing remarkable development in many directions. But consider in particular some of the discoveries that bear on the
possibility of changing humanity. We now understand the structure of the
material which carries most of our genetic inheritance, DNA, and we have
the code by which the amino acids, the constituents of proteins, are made.
Thus there is the possibility of direct chemical modification of inheritance. Human tissue has been grown and cultivated in the test tube far
past its natural lifetime. We have discovered how to produce hybrid cells
between diverse vertebrate animals including a hybrid between man and
mouse, and these hybrid mouse-man cells reproduce themselves. We
commonly stock sperm of livestock in deepfreeze, the sperm bank producing desirable and viable offspring. One can foresee the possibility of a
bank of eggs as well. We know how to regulate the sex of the offspring of
rabbits by removal and replanting of an ovum. We have discovered a good
deal about the defenses of the body, viral invasion and protection of cells,
and many other powerful regulatory mechanisms of life. Surgeons can
transplant important organs from one person to another, and thus bring
into some question the integrity of the individual. An increasing number
of chemicals which strongly affect the state of mind, mental health, or
mental abnormality, are becoming familiar.

These things have their shocking aspects. They, and the possibility of
sudden death from nuclear bombs, the difficulties within our cities and so-

cial maladjustments which technology has brought, make some feel that we should turn back from science. Do we have enough science, and should we halt science or turn it back while we stop and think? We certainly should think. But turning back from science is a different question. It is essentially impossible. This would be like deciding that breathing oxygen is not a good idea and that we should return to the waters from which our fish ancestors came, since it must have been a happy existence. But such is simply not the nature of man. Curiosity, knowledge, and understanding represent some of our most important God-given gifts; they are characteristic of man and part of his nature. The discovery of new things represents one of man's most satisfying achievements and knowledge one of his most impressive and permanent monuments. Any who try to turn away from scientific understanding are like fish out of water. They will be overtaken by others more attuned to man's inheritance and future. Inevitably we are faced with exciting knowledge, enticing knowledge; yet we are also faced with very hard and demanding decisions—demanding of all wisdom that we may have. This is man's glory and his challenge.

Civilization needs people who know how to do things. But it needs most those who know what to do. And in the effort to properly use our resources and knowledge, every citizen must be a participant. Everyone in every walk of life has a responsibility to understand science and its meaning, for he must understand where civilization is going and where it can go. The citizen needs to be scientifically literate. And every scientist needs to have a breadth of understanding and a sense of values which qualify him to deal with the human role of science.

Research Labs: Variety and Competition

A variety of institutions, with differences in style, facilities, personnel policies, motivation, and points of view is important to research because of the enormous variety in the paths to discovery. Discovery is, of course, inherently unpredictable. No one approach, regardless of how clearly it may illuminate some facets of knowledge, is likely to reveal all aspects even of a small field of science. Differences in style and purpose within and among university, industrial, government, or private research laboratories therefore represent more nearly an essential variety than several imperfect approximations to the perfect laboratory.

Yet these various research efforts overlap in important ways. They compete for outstanding personnel, for funds, and for research success in those areas where their efforts are applied to similar problems. Like variety, such competition is useful, as long as it does not wrest priority from efficient work toward the primary task of discovery. In this competition, there is a dynamic equilibrium among institutions which is reflected in the quality and number of their personnel, their available resources, and their research successes. Some aspects of this equilibrium have lasted a very long time. Others can easily shift within a period of a few years. Differentiation and competition are discussed here not in order to enhance competition, but in order to focus on a few strengths and weaknesses of different types of institutions with the hope of discovering useful modifications and adaptations.

"Dynamic equilibrium" is necessarily an inexact description for what exists among laboratories and types of institutions. While personnel and funds can, in principle, shift from one type of laboratory to another within a year, in fact national policies and time scales are such that such changes normally take at least several years. Furthermore, most research is of such long-term nature that about five years is a minimum period for establishing a measure of accomplishment or failure. The driving force of national needs and policies, and the results in accomplished research, have a similar time scale for change, although they are often slightly out of phase.

From time to time, the United States as a whole may face a more rapid increase in the supply of research-oriented personnel than in the number of positions and apparent need for such people. It may also experience change in the availability of financial support for research. These shifting conditions naturally cause readjustments in the optimum policies and relative successes of institutions. They also induce shifts of student enrollment in various research areas, shifts which may, by being out of phase with fluctuating demand, build the manpower disparities for the future.

Institutions engaged in research can be divided into four principal types: universities, industrial laboratories, government laboratories, and privately supported research institutions. In the category of government laboratory, I intend to include both laboratories run directly by a government agency and large centralized laboratories of unified purpose supported entirely by government funds, but run by some other agent. Under privately supported research institutions, I include those privately supported institutions which do research but essentially no classroom teaching. I would like first to compare three broad aspects of the various laboratories: their roles and goals, their facilities, and their personnel policies and attitudes.

The primary roles of the university are to teach and to develop new knowledge and insights. Provision of the highest levels of training in a given field almost necessarily leads to a concomitant development of new knowledge and insights. Students are exposed to the forefront of information and thought, and trained in developing new knowledge and insights themselves. Hence the teaching role of the university leads very naturally to a strong emphasis on basic research.

I will use the word "basic" to refer to work not directed toward an immediate practical application. The emphasis here must be on the word "immediate," since almost all basic work finds practical applications on some time scale and possible applications are always in the background. What has been called "strategic" research is basic work in an area judged to be useful for some specific purpose; hence it is directed toward practical applications, but not on any immediate or predictable time scale. The convenience and widespread use of the term "basic" does not mean that any sharp separation can be made between basic and applied research. Very frequently both motivation and results in a given project are mixed or intermediate in character, and often the support of research is motivated by interest in its applications but the work itself is carried out by people with attitudes which are basic in character.

Only private research institutions can have a commitment to basic research comparable to that of universities. Such institutions and their purposes are highly variable. Some of them, such as the Carnegie Institution of Washington or the former Rockefeller Institute of New York have been even more specifically directed toward basic research as a primary function than any university.

The role of industry in basic research is much more ambivalent. Its goal of producing a product or service effectively and cheaply naturally generates support for product development and applied research. The route from basic research to a company's product is usually long and chancy. Furthermore, even when new research knowledge is needed to develop a product, an industrial firm might obtain the information it needs most efficiently and cheaply from published literature or from others whose primary role is basic research.

In the long run, however, the value of research to industry or to the economy as a whole is often overwhelming, especially in industries where technique and technology can change rapidly. Consider long-distance telephone communications. Between 1935 and 1965, the cost of a long-distance telephone call across the North American continent decreased by a factor of about four in actual dollars. In view of inflation, this was a reduction in real cost by a factor of about twelve, and since 1965 direct dial costs have decreased still further. While some of this cost reduction surely resulted from increased use of the telephone, much of it derived from research and development, some of it quite basic. A very large fraction of the applicable research and development took place in the Bell Telephone Laboratories, the dominant communications laboratory in the United States, a fact which enables one to specify its character and cost with some surety. The yearly bill in 1965 for all long-distance telephone calls in the Unites States was about $4 billion. With the knowledge and technology available in the mid-30's, the same service would probably have cost 25 billion dollars or more. For such service, present technology thus saves the American public many billions of dollars each year. The regulation of rates for the telephone system has also meant that steady reduction in cost of service provided one of the few routes for the company to increase its profits. How much did this research and development cost? The entire telephone system support of the Bell Telephone Laboratories from its beginning in the 1920's until 1965 was somewhat less than $2 billion, Thus, the decrease in cost of services each year is much larger than the total monetary support of the Laboratories over their entire history, and these contributions to long-distance telephony are, of course, only a part of the Laboratories' work. Among others, the chemical industry can produce parallel examples of the value of research and development.

In spite of such overwhelming figures in certain specific cases, for an individual company the decision whether and how much to support long-range research remains difficult. Because of the rapid diffusion of knowledge and the relatively small cost of patent rights, knowledge and technology can often be adequately borrowed or otherwise obtained more cheaply than by direct support of research. Even on a national scale, how much large-scale investment to make in basic research is not always clear, since one nation can profit from discoveries in another without having to make heavy investments in research itself. However, clearly any very large nation or corporation which hopes to be on the forefront of technology and effectiveness should make substantial investments in long-range research; a leading organization can hardly borrow most of its knowledge from those which do not lead.

In the industrial context, appropriate basic research can be expected to be of a largely "strategic" nature. Even about supporting basic research of a strategic nature, however, corporations often manifest ambivalence and vacillation. Basic research may be initiated in an over-enthusiastic hope of near-term applications, or in the belief that research will improve the corporation's image, or to induce highly trained personnel to associate with the corporation. Such conditional support is likely to be undermined whenever an industry is under economic pressure or undergoes a change of management. Unrealistic orientation of research programs coupled with short-term support has produced many disappointments both for management and for research personnel in industrial organizations.

The role and aims of government laboratories are rather more like those of industrial laboratories than of the universities. However, profits are invisible, competition much less in evidence than in industry, and stability usually greater. Consequently, longer-term and higher-risk research, of both basic and applied nature, are frequently supported in government laboratories. Unfortunately, the paucity of competition and the lack of profit accounting can also encourage tolerance of mediocrity and unproductiveness.

Assuming adequate funds are available, the privately supported research institution is free to adopt whatever goals and character its supporters and personnel may wish. When well financed and intelligently manned, it can have the great advantage of not being accountable to anyone else. Capable and lively personnel in private research institutions often create the conditions required for very outstanding success in basic research, but sometime they suffer from lack of contact with problems of the world and with the flux of ideas.

As I have noted, the goals and roles of industrial and governmental laboratories are frequently congenial to applied research and to development. In these areas universities and many privately endowed laboratories are at a distinct disadvantage, because they are not so faced with specific applied problems to be solved, and often do not have access to the large-scale facilities that are often important to applied research.

Research facilities of institutions may be described in terms of equipment, working space, and technicians or assistants. One overall measure of the cost of facilities and abundance of assistance is the overhead rate on salaries of research personnel, since it is this overhead which pays for most of their cost. By this measure, universities and industrial laboratories are in sharp contrast. The characteristic overhead rate in universities is about 50% of salaries, whereas that in industry is 100% to 200%. In this respect government laboratories are more comparable with industry, and private research institutions, while quite variable, are often more comparable with universities. Thus university research laboratories tend to be skimpy on technicians, services, and general facilities. Industrial laboratories are often outstandingly equipped as far as general facilities and certain kinds of assistance are concerned, although high-quality technicians are generally in scare supply and hence not numerous even in many well-financed industrial laboratories.

Government laboratories also tend to have large amounts of equipment and a good number of supporting personnel. Often they have impressive investments in large-scale costly apparatus and equipment. Some of this may be unique, well planned, and well used, but in too many cases a piece of expensive equipment is poorly planned and manned, and comes uncomfortably close to being a white elephant. In certain cases, the primary function of a government laboratory is to build and operate some unique set of equipment to be available to the scientific and technical community at large. In other cases, an expensive or unique piece of equipment may be built apparently with the primary motive of upgrading the standing of the laboratory and attracting high-quality personnel. This phenomenon also occurs at times in industrial laboratories.

The lack of technicians and supporting personnel in universities is compensated at least partially by the supply of students who, while they may not initially be very expert, are likely to be bright and ambitious and to find apparatus construction and other chores involved in research very useful as a learning mechanism. In many fields, however, recent trends have decreased the number of research students, and may eventually result in inadequate manning of university research.

In the area of personnel policies and attitudes there are marked differences among the various categories of laboratories. These differences affect the attitudes and careers of individuals who populate laboratories, and also strongly influence them in their selection of a given setting for their work.

The dominant aspect of life in a university is commitment to teaching, which for research implies the flow of students and other young scholars in and out of the university laboratory. Both classroom and informal personal teaching distract the academic staff of the university from its research function. On the other hand, they provide a continual challenge of ideas and a demand for breadth which result in a research atmosphere of unique quality. The earlier stages of the careers of academic faculty are marked by a competition for recognition and favorable positions which may be as intense as the rather more publicized competition for executive positions in the industrial world. Later, the academic researcher tends to continue this habit of competition but without such intensity or any immediate visible effect on his career. The primary goals in this competition are research accomplishment, freedom and facilities for research and only occasionally administrative position. While the choice of research activities is strongly affected by the availability of facilities and of funds, there is seldom any university or departmental policy which directs the research of an individual faculty member. He is usually happily, but sometimes starkly, on his own.

In an industrial organization an individual scientist may, in some cases, have substantial freedom to direct his research, but at the very least he is expected to see that it is attuned to the needs of the corporation; in other cases, he finds himself closely directed by management to undertake a specific investigation, the nature of which may be changed on short notice. Much of the power and effectiveness of the industrial laboratory, in applied research at least, is connected with just this ability to direct a team of scientists and engineers and to enlist their attention and cooperation for intense work on a specific problem. It is a directed teamwork which is possible in universities only under very unusual circumstances, such as war-time emergency. This very process, however, can be quite disruptive of sustained basic research and scholarly effort, and industrial organizations have considerable difficulty in maintaining both the strength of directed teamwork and the depth of scholarly approach required for outstanding basic research. In the industrial environment, frequently the primary competition is for administrative position, since it involves the authority to direct the efforts of a large team, or perhaps of many teams.

Such laboratories have severe difficulty finding ways to properly recognize and reward a research person who may be outstanding in his field and creative, but a lone wolf with no wish nor suitability to move upwards in organizational and administrative responsibility.

An industrial laboratory which recognizes that it has a continuing need for knowledge and expertise in a particular field may be willing to support someone to devote a substantial part of his life to this area. His long-term commitment to this field can be valuable in that he has time to develop deep and detailed knowledge. Furthermore, he has a natural and frequent stimulus from applied problems. On the other hand, without teaching and the challenges of students and a variety of colleagues, or under proprietary pressures from his laboratory to maintain a certain distance from competitors, such an individual can become quite specialized and inflexible, missing opportunities to synthesize knowledge of his own field with that of others.

Personnel policies and attitudes in government laboratories vary widely, but are frequently intermediate between those of industry and those of the university. The competitive pressure typical of industry are not so omnipresent, so that long-term programs with rather stable goals and directions are common. Except concerning work to which access is restricted for military reasons, government laboratories are characteristically quite open and untroubled by proprietary restrictions on their communication with others. They may even be forced into frequent contact with other scientific groups if their function involves service to industry or to universities. While civil service policies favor administrative roles for able individuals, the difference in salary and standing between the successful individual research person and the manager is not as great in government laboratories as in industry. Frequently a government agency is quite protective of its laboratories and gives employees there advantages in the availability of equipment and services which can be quite attractive to them. Thus, a stable government agency which establishes enlightened policies may provide a good climate for long-range and intensive research and thus produce very successful, steady, and systematic programs. On the other hand, the lack of intense competition characteristic of the early stages of academic life or of industry can also result in slow progress and mediocrity.

Private research institutions are so individual in their characteristics that their personnel policies and attitudes are difficult to describe with any generality. However, at least one characteristic is common to many of them. The most outstanding success of a private institution often occurs within a generation after it is first organized. If the institution is well endowed and its policies intelligently planned, it may be successful in obtaining initially very talented research personnel, both leaders in their prime and promising young people. These individuals are likely to make the institution particularly effective during their own active careers. However, once the institution has hired to the saturation point, it may gradually lose contact with active young people and find itself with an aging and somewhat stagnant research corps. Thus, without a substantial connection to university life or to active industry, and unless some major reorganization and rejuvenation occurs, such laboratories may find their most successful period lasts only about one generation.

My comments and generalizations are inadequate, of course, to represent all laboratories, some of which may have characteristics determined primarily by the genius of one or a few individuals, or by some particular setting. A part of a given laboratory may also blossom extraordinarily or be completely atypical because of individual talent or special historical developments. Yet I hope I have properly reflected a few important and common properties of research institutions. It is important to recognize that some of these properties, good or bad, are inherent in the type of laboratory and in its social and organizational context. However, efforts to capitalize on their better features and to ameliorate their inadequacies might well improve the productivity of each type of laboratory. Consider now some of the steps which may be indicated.

A measure which immediately comes to mind is that each type of laboratory might try to adopt the good features of others. One of the most successful ways of accomplishing this is by association; the carefully planned association of different styles of laboratories should be viewed as healthy symbiosis rather than as an unnatural marriage. For example, a university associated with a government of industrial laboratory may acquire thereby the stimulation of constant contact with applied problems, and also the large-scale facilities often necessary for applied research. Correspondingly the government or industrial laboratory acquires increased contact with talented young persons and the openness of a university environment. In some situations, such associations work well. In others, unsatisfactory arrangements or inequities in the strength of the two associated institutions may enable the stronger institution to dominate and injure the weaker one or may obviate any substantial influence on and benefit to the stronger institution. The point of an association is to make available some of the helpful natural attributes of each institution without destroying the character of either. Privately supported research laboratories are flexible enough in character and goals to find association with universities particularly easy, they help to provide the university with facilities and outstanding personnel while benefitting from the breadth of interests and the flow of young people characteristic of the university. Very successful associations of government laboratories with universities are represented by the Lincoln Laboratories of M.I.T. and the Joint Institute of Laboratory Astrophysics at Boulder. Satisfactory close association between an industrial laboratory and a university laboratory is much more difficult to achieve, since it tends either to decrease the openness of the university or to amount essentially to a subsidy of the university by the industrial unit, a subsidy which an individual corporation cannot long support.

The need of university laboratories, particularly those doing applied work, for contact with real problems and with large on-going operations, can be partially solved by some other types of very loose association. One is the consulting for industry or government often done by university personnel. In applied work, or to maintain general breadth of experience within the university, this is a valuable and valid association. It should be cultivated, but with some care to see that it is of real value to the intellectual growth of the university staff and not so engrossing that it detracts from their devotion to educational work. Another partial solution which might be considered a distant association is the planned temporary exchange of personnel, and even the competitive hiring of research personnel away from one laboratory into another, with the resulting transfer of attitudes and experience. Research administrators would do well to find ways of facilitating useful exchanges of personnel for terms as long as a number of months, or a year.

University scientists had strong contacts with applied problems and with industrial laboratories during World War II when universities devoted much of their effort to military developments. This interaction between different types of laboratories and research personnel interrupted basic research, but it has also had important beneficial effects on research and research personnel in all U.S. laboratories involved in the physical sciences. Perhaps another substantial contact of university personnel with applied work is developing at present as the result of interest in environmental problems. This is an area which will need for a while the open-ended and free-wheeling university approach as well as systematic long-range and detailed work which government laboratories can provide, and the directed, practical drive of industry. If national interest in environmental problems turns out to be long-term and substantial, it could be important both as a stimulus for new research and fresh institutional connections, and as an orientation for education.

A second general method for adapting the better features of one type of institution to the needs of another is direct adoption. Industry has, in some cases, simply tried to set up the motivational content of a university within its own house, giving relatively free rein to scientists for basic research and choice of problems. In cases where this has been tried in rather pure form, it has usually failed because the efforts and loyalties of research personnel became quite unrelated to the industry, and could not be permanently supported. In less extreme forms, it is successful in many cases—so successful that universities have sometimes found industrial laboratories difficult to compete with for the most outstanding talent.

It may seem peculiar that an increase in the effectiveness of one type of laboratory can be harmful to others, but competition makes this true. Hence at times competition needs to be thoughtfully compensated for on a national basis, or for that matter on an international basis when the research environment in one nation advances too far ahead of that in others. Such a situation arose, for example, following discoveries in solid state electronics during the 1950s. Outstanding basic research in solid state physics came to be so highly valued in industry that the most productive scientists trained in solid state work were provided not only with excellent facilities and relatively high salaries in many industrial companies, but also with the quality of research freedom generally associated with universities, without the distractions of teaching or administration. The resulting competition for qualified personnel was so difficult for the universities that it posed a threat to continued top quality solid state research in academic institutions and hence to education in this field. The threat was ameliorated and reasonable competitive balance restored by the founding in key universities of special government-sponsored materials laboratories with needed facilities and support. In other branches of electronics, industry had from the beginning taken such a commanding lead in research facilities and opportunities that few universities could compete for the most outstanding personnel.

Privately supported research institutions have in many cases very successfully rejuvenated or kept fresh their research life by association with a university, by direct adoption of the practice of teaching within their own institutions, or by some combination of the two. Interesting examples are the Carnegie Institute of Washington, the Woods Hole Oceanographic Institution and the Rockefeller Institute, now Rockefeller University.

Universities have made substantial efforts to adopt at least one favorable feature of government and industrial laboratories by improving facilities. This has its value, but it has also encouraged a tendency of university research to move toward the use of large centralized equipment and shops. While this move has elements of efficiency and may be appropriate for other types of laboratory, it can weaken the university's teaching function by depriving students of firsthand experience with all elements of the research process. It is important for universities to resist overcentralization or overdependence on the centralized operation of large-scale equipment. Some types of research, of course, demand large-scale facilities. Too often, however, in work on large centralized equipment, the student is either engaged only in a minor facet of the entire enterprise; or he has the apparent privilege but the educational disadvantage of a user who requests that certain experiments be done and is then presented with the results, a process which allows him only second-hand contacts with important aspects of apparatus design, debugging, and operations.

University research could be facilitated by additional support personnel. Furthermore, shop and technical personnel closely associated with smallish research groups and available to interact intimately with students can be very effective in familiarizing students with basic techniques.

At present, university departments are typically populated by a combination of students, young Ph.D.s, and faculty members primarily oriented toward their own personal research. Too few supporting and intermediate research personnel stay permanently in the university, although they could be invaluable in maintaining continuity of programs and skills and in giving assistance to students in technical or specialized aspects of research. If the number of advanced science students continues to decrease, such personnel would be especially important in maintaining successful ongoing research programs, directed by the teaching staff. Such supporting technical personnel could serve effectively on a part-time basis. Hence positions of this type may be appropriate for women with young families, providing them with continued contact with front-rank research.

Industry needs first of all to find ways of establishing clearly and consistently just where its interests in research lie. Hopefully, once it does so, the large corporations at least will give stable support to strategic as well as to applied research. Most of the smaller corporations should probably not maintain substantial research efforts at all. They do, however, need clever work on product development, which might sometimes be labeled research. Such work can in fact be the lifeblood of a small technologically-oriented company. However, uncertainties and fluctuations in the rewards of longer-range investigation may be too great to justify its consistent support within a small company; furthermore, the disadvantages both of small size and these uncertainties tend to undermine the effectiveness of whatever modest long-range research is undertaken. Larger corporations can accept risks and fluctuations in the research game, and indeed those which use technology extensively must maintain successful and alert research laboratories if they are not to become antiquated and noncompetitive. Can this give them too dominant an advantage over smaller corporations? Yes, in many cases it could in the absence of pubic concern and of adequate sharing of technology.

Industrial laboratories must seek personnel policies which recognize the value of individual senior scientists who do not undertake management as a career, and which allow them to participate actively in the scientific community. In universities, a department chairman frequently works energetically to see that salaries of outstanding professors within his department are increased far beyond what he expects to receive himself. By contrast, the traditions and relatively structured life of an industrial laboratory make it psychologically and practically difficult for an administrative boss to see that an outstanding research person under his direction gets a salary consistently greater than his own. But such a phenomenon is not impossible, and might provide some needed recognition for creative research personnel. Recognition by his peers is probably still more important than a large salary so that giving him opportunities to mix easily with the national and international community of scientists and to publish freely may be the best way to recognize him, and at the same time, to provide the contacts and criticism needed to keep the industrial laboratory alert. Exchange of personnel with other types of institutions, particularly universities, can also provide some of these opportunities and breadth to industrial laboratories.

On the other hand, we should not expect that the industrial scientist will characteristically be a generalist in the same sense as a university scientist and teacher tends to be. Instead, if he is consistently supported by industry, the industrial scientist can have an unusual opportunity to cultivate deeply a particular field and to become outstandingly expert. While the management of the industrial laboratory must maintain its effectiveness in coordinating intensive efforts of teams of research personnel, particularly in the area of applied research, it must also find ways to maintain research of a different quality, of the longer-range and more individualistic variety which requires at least partial insulation from the immediacy of daily demands and shifting crises of industrial life.

Although industrial research laboratories differ widely in quality and in their attractiveness to outstanding research personnel, for the most part they have had difficulty competing with the more prominent university and private research institutions for high-quality research personnel oriented toward basic work. Periodic slow-downs in university hiring, mostly the result of a normal growth and maturation cycle, create a job shortage which, though difficult for such personnel, may benefit other kinds of research institutions by providing a better balance in availability of highly qualified personnel, thereby creating an important opportunity for other laboratories with suitable openings and research policies.

Government laboratories serve a number of unique functions. In some cases, laboratories are created to make needed attacks on problems whose national importance has rather suddenly become evident. A centralized government laboratory can be built quickly to focus research on such problems. In other cases, laboratories must be maintained in areas where an agency has an important interest and needs its own experts both to carry out some of the research involved, and properly to review, oversee, or even direct efforts of other laboratories. In still other cases, central facilities are needed either for a research area of importance to a government agency or for the general use of the scientific community. However, the large resources sometimes available to government laboratories and the frequent lack of real competition in research performance are special hazards for them.

Industry normally competes with universities for personnel by offering high salaries, good facilities, and possibly administrative power; government laboratories, however, generally offer competition to universities only on the last two scores. Agencies of the government frequently gravitate in the direction of maintaining unnecessarily expensive and not particularly productive equipment. In many cases, the difference is enormous between the supporting facilities available to an individual in a government laboratory and those normally available to him in a first-class university; this disparity exists even in laboratories operated for the government by universities. Perhaps some such differences are essential if the government laboratories are to continue to exist and carry out their functions. However, in many cases such differences are so large that they are probably not tolerable over long periods of time.

A government agency is naturally concerned about the fate of one of its own laboratories, and may hence be overindulgent. This instinct is particularly hazardous when the same agency, perhaps even the same individual, is responsible for initiating or maintaining the laboratory, as well as for making decisions on competing bids by other institutions for research support. There are of course valid reasons for some laboratories to have much better facilities than others and different styles of operation. However, some of this unhealthy disparity often evident between government laboratories and others could probably be avoided by administrative care to remove the decisions on granting agency funds for supporting research from those who have a special and personal interest in the maintenance of one of the agency's own research laboratories.

Like a private research institution, a government laboratory may be very effective while young, but become stagnant sometime later, particularly if the problems for which it was initiated have become less urgent than when they first attracted attention. Despite political and human difficulties, support should be removed from those laboratories whose mission has been completed or has disappeared and which cannot be effectively reoriented to work on current vital problems. One of the more successful government laboratories, the Naval Research Laboratory, has been in operation a long time. In addition to the continuing viability of its mission, its success seems closely related to maintenance of high quality basic research and close contact and competition with the rest of the scientific community. Methods must be found to enhance such contacts, particularly for government research centers not already closely associated with universities, and to challenge continually the efficiency and success of government laboratories.

The decreasing availability of university positions and a growing tradition in government laboratories of employing young postdoctoral personnel for a short period after they receive the Ph.D. degree seem to offer real promise for maintaining a flow of younger people through such laboratories, and for giving them experience with the values and opportunities there.

Clearly, government laboratories should cultivate their unique roles and advantages. They can serve well, I believe, in any long-term systematic work of either specialized or broad interest to science which is not carried out by industry. Such long-term systematic work is increasingly difficult to maintain in universities. A steadily growing need of science and technology is for the critical collection and codification of information in the many fields which have been advancing so rapidly during the period of vigorous growth and enthusiasm for science over the last several decades. While university personnel may be the most obvious source of textbooks, monographs and critical tables fit naturally into the roles of many government laboratories. The effort needed to digest information, carry out systematic series of measurements, and make the results better available for use is large, but it could be of enormous value to further progress everywhere.

There should be special concern in government circles and in the technical community for the health and efficiency of the laboratories of the Department of Defense. Some of these are indeed on the forefront of research and some are working away very effectively on longstanding problems, more or less unnoticed by much of the technical community. But too many have long suffered from stagnation and mediocrity, a fact which, coupled with some isolation due to military classification and the disinterest of much of the scientific community in perfectly real and valid military problems, may in coming years make these laboratories particularly ineffective and wasteful. Both the administration of the Department of Defense and the technical community have a responsibility to find ways of eliminating some and refreshing other such laboratories.

The ingredients of a successful research environment, and the difficulties of research administration are clearly even more complex than my already involved discussion indicates. Two more factors must at least be mentioned: the crucial role an individual can play, and the powerful effect of tradition. These factors, though not simple to analyze, are easily illustrated—the first strikingly so by the fructifying effect of Enrico Fermi on Italian physics and physicists, or of Ernest Lawrence on the University of California. The second factor, tradition, creates an expectation and ambiance which are often valuable but slow to change either for better or worse. To illustrate this I shall be quite personal. One of the clinching arguments for me when I accepted my first job, at the Bell Telephone Laboratories, was that Davisson and Germer had done their famous experiment there. After I transferred, a decade later, to Columbia University where I. I. Rabi's influence was then enormous, I found still a few bright-eyed young men who had been stimulated to come to Columbia by the interesting figure of Michael Pupin, dead more than a decade and long before that finished with any active role in science.

SPIRITUAL VIEWS FROM A SCIENTIFIC BASE

How and Why Did It All Begin?

Two Views of Creation

There have been two distinct, long-standing views of creation—that is, the origin of our universe and of life in it. The first view makes each the result of a unique and special event. The second assumes that the universe and life are more continuous and commonplace events, with no unique moment of origin. These two possibilities do not represent simply the difference between a religious view and a secular view, though they may affect profoundly an individual's outlook. Nor do they correspond to a scientific versus a non-scientific view. Assumptions of a unique and special creation on the one hand, or an inevitable and more commonplace one on the other, have been a continuing theme through a great deal of man's thought.

During much of human history, it has been normal to believe that life was created from time to time spontaneously from materials of nature, or perhaps by some more or less capricious supernatural event. A recipe for creating mice during the Middle Ages advises taking an old shirt and putting some grain in it. When stuffed into the corner of a room for a few weeks, mice were sure to be found, created according to the recipe. Such ideas as the commonplace spontaneous creation of life persisted into the nineteenth century and were only disproved after much labor and considerable argument within the scientific community by the great scientist Pasteur. By aseptic techniques, he made a convincing case that all life as we know if comes from other life.

While science was thus on the one hand making the creation of life seem rather special, on the other hand it was also busy during the same period detracting from ideas about the unique character of man's existence. Copernicus and Galileo had already removed man from the center of the universe. The study of the vast collections of stars called galaxies, and then of cosmology, extended our view of the universe so enormously that man's being important in it seemed almost unthinkable. Darwin's ideas on evolution, and now modern biochemistry, go a long way towards indicating that life itself was generated by random processes, some might say rather casually and accidentally, on the basis of physical laws which we largely know.

Random Creation

This general view of random creation is, however, by no means a product of recent thought. Lucretius, the Roman poet and a proponent of an atomic theory of matter, made the following remarkably modern-sounding statement more than 2000 years ago:

> Our world has been made by nature through the spontaneous and casual collision and the random and purposeless congregation and coalescense of atoms where combinations could serve on each occasion as the starting point of substantial constructions—earth and sea and sky and the races of living creatures. You have the same natural force to congregate them in any place precisely as they have been congregated here. You are bound, therefore, to acknowledge that in other regions there are other earths and various races of men and breeds of beasts.

I believe it was Julian Huxley who first used the example of a hundred monkeys pecking randomly at a hundred typewriters in order to suggest the randomness and lack of mystery even in man's intelligence. He noted that the monkeys would in time, entirely by chance, type out all of Shakespeare's works and *The Encyclopedia Brittanica*.

These ideas are certainly cogent to our problem. However, to put this randomness in a little more perspective, we must note the results of quantitative calculation. While it is true that monkeys may randomly turn out *The Encyclopedia Brittanica*, a simple calculation shows that one billion monkeys typing randomly as fast as they can 24 hours a day on one billion typewriters for the entire lifetime of the universe as we know it would probably not yet have typed out the correct sequence of letters in the title *The Encyclopedia Brittanica*. Thus, while randomness must have had an important and powerful effect, something other than the simple random juxtaposition of atoms must have been important in the formation of complex life. We seem to need something more systematic, some mold from which the complex patterns of creation could develop. The scientist would assume these patterns have been guided by aspects of the laws of physics and chemistry which we simply have not yet quite grasped; others may assume the hand of God. As our insight becomes more penetrating, how different will these two views really seem?

Insights from Astronomy

Much of the modest amount we know as scientists about our origins comes from astronomy, as guessed by Alexander Pope when he wrote, somewhat over-hopefully, of the astronomer:

> He who through vast immensity can pierce
> See worlds on worlds compose one universe
> Observe how system into system runs
> What other planets circle other suns
> What varied being peoples every star
> May tell why Heav'n has made us as we are.

In the 1960s there was a remarkable discovery of microwaves—that is, short radio-like waves—which uniformly pervade all space. We can presently understand their existence only if they represent radiation left over for us from an initial enormous explosion of the universe. This radiation, more than any other one piece of evidence, seems to lead inevitably to the conclusion that the universe did indeed have a unique moment when it was small, enormously hot, and expanding rapidly—the so called "big bang." Some scientists still doubt such a conclusion, and continue to look for an explanation in terms of an ever-existing, never changing universe. But so far they have been unsuccessful. The microwave radiation we now see must have been created during the first one hundredth of 1% of the lifetime of our universe—a lifetime which from this origin until now must be about fifteen billion years. Thus we have remarkable scientific proof that there was indeed a unique moment in the creation of the universe. In addition, our most powerful telescopes seem recently to have penetrated far enough into our universe to approach its boundaries, and catch a glimpse of how it looked when much younger.

Why all this lapse of time from the origin of the universe, about fifteen billion years ago, until the creation of man, whose existence on the earth surely isn't much older than a few million years? Are we a random afterthought? Hardly that, for we understand now that before complex life could be created, materials of the universe had to be properly cooked and processed. Stars were formed, and went through their cycle of billions of years of life until, with a majestic display, they exploded and spewed out the heavy chemical elements it was their destiny to produce from the materials available in the new-born universe. Elements which they emitted were gathered together into new stars, the so-called second generation stars of which our sun is one. Thus the sun and its satellite the earth can contain some of the needed heavy chemical elements such as iron for blood, calcium for bones, and iodine for metabolic chemistry without which our life would be difficult to imagine. Just these preparatory processes would require, from the nature of physical laws they followed, almost half the life-span of the universe.

About four and a half billion years ago, shortly after the formation of our second, or possibly third-generation star which is the sun, materials of the earth solidified. One and a half billion years later, that is about three billion years ago, life began on it and we can trace from that time its steady and fairly orderly development.

Are We Alone?

Was this development, eventually producing man, peculiar and unique? Are we alone in the universe, or is our planet one among billions which support sensitive and intelligent life? The total number of stars in our galaxy, each of which might possibly support life around it, is about one hundred billion. But ours is only one of ten billion such galaxies within the universe. Hence, with one hundred billion times ten billions of different stars within the universe, it is natural to conclude that our existence is insignificant, and that life must have developed myriads of times, with some forms much superior to our own. However, as in the case of the monkeys typing randomly, something more may have been needed than just all those random chances.

We do not know for sure just how planets are formed, nor hence the chance of a star having a planet such as ours. Recently geophysicists have discovered that there was an enormous stellar explosion in the immediate vicinity of our star the sun just before the planets were formed. Is some special circumstance like this required? The nature of a planet on which life can begin clearly is rather specific and circumscribed. How likely is it that conditions as favorable as those on earth occur in other planets? If there did happen to be a planet of the right qualities, would appropriate molecules inevitably come together to form the complex assemblages which life seems to require? What is the nature of the step from apes to man, producing a mind which conceives of astronomy, or of studying its own origins? We know a great deal and yet little of such matters.

Biochemists have made convincing arguments about what kinds of molecules might initiate the life processes. Radio astronomers have learned that all of the simple molecules which biochemists believe are needed for a start in the process of building life—all of those needed for the reproduction of the simplest polypeptides or protein-like substances—can be found in dust clouds in interstellar space, even before these clouds gather into stars and planets. But now, given these materials, scientists are still groping to see how they might have built up complex forms needed for reproductive life.

What definite hope can we have of knowing whether we are alone, or our civilization is repeated and surpassed billions of times among the stars? Even the possibilities of knowing are impossible to state, because the most important scientific discoveries are frequently unimagined until they are discovered and surprise us. For the moment, our best hope of knowledge of other life is to leave the laboratory and go exploring. Some civilization, perhaps only a few hundred years more advanced than ours, might have already guessed at our existence and be trying to signal us. What would it mean to man's perspective if suddenly we received messages and wisdom from other worlds? We've listened very carefully, guessing what kind of signal might be used, and so far heard nothing.

Advances in Space Work

Advances in space work have considerably enlarged our explorations. It has been commonly thought that on some of our sister planets, such as Mars, Venus, or even the Moon in an early state when it might have had an atmosphere, other life could exist. While the Moon now has no atmosphere, it has seemed to be marked by rills and valleys which are difficult to explain except as due to a running fluid, such as water, sometime in the past. Unfortunately, while our explorations there continue to intrigue us with information about the early history of the moon and the solar system, the Apollo flights have shown that water and organic materials are rare enough on the lunar surface to dash most hopes of finding traces even of past life there. Recent measurements also show that the surface of Venus is overwhelmingly hot: 600° Fahrenheit and far above the boiling point of water. Such temperatures are quite inimical to any form of life we can presently imagine. Could there be life possibly in the cooler upper atmosphere of its clouds? Just possibly. Jupiter, Saturn, and the other outer planets are generally too cold to be likely supporters of life, Mercury too close to the sun and too hot.

Thus, scientific discoveries show us clearly that, at least within our solar system of nine planets, our earth is truly a gem, and its life unique. We know further that it will at least be a long time before man directly encounters any extraterrestial creature remotely like himself; perhaps, contrary to many generations of fiction and to the common expectations of many scientists in recent decades, we are indeed alone and unique in our universe.

How our developing scientific understanding either changes or reinforces religious views is a question each individual will answer for himself. Yet any substantial success in the common search by religious or scientific approaches for the origins and meaning of life must inevitably mold man's view of himself, and recent recognition of the special character of this planet and its life can only heighten man's awe. For the future, human thought and instincts, the innate creators of science, will surely lead us further in exploring our origins and towards understanding man's remarkable situation. If such understanding substantially increases our sensitivity to the wonders we see, and to the sacredness of life, it will serve us well.

The Convergence of Science and Religion

The ever-increasing success of science has posed many challenges and conflicts for religion—conflicts which are resolved in individual lives in a variety of ways. Some accept both religion and science as dealing with quite different matters by different methods, and thus separate them so widely in their thinking that no direct confrontation is possible. Some repair rather completely to the camp of science or of religion and regard the other as ultimately of little importance, if not downright harmful. To me science and religion are both universal, and basically very similar. In fact, to make the argument clear, I should like to adopt the rather extreme point of view that their differences are largely superficial, and that the two become almost indistinguishable if we look at the real nature of each. It is perhaps science whose real nature is the less obvious, because of its blinding superficial successes. To explain this, and to give perspective to the non-scientists, we must consider a bit of the history and development of science.

The march of science during the eighteenth and nineteenth centuries produced enormous confidence in its success and generality. One field after another fell before the objective inquiry, experimental approach, and the logic of science. Scientific laws appeared to take on an absolute quality, and it was very easy to be convinced that science in time would explain everything. This was the time when Laplace could say that if he knew the position and velocity of every particle in the universe, and could calculate sufficiently well, he would then predict the entire future. Laplace was only expressing the evident experience of his time, that the success and precision of scientific laws had changed determinism from a speculative argument to one which seemed inescapable. This was the time when the devout Pasteur, asked how he as a scientist could be religious, simply replied that his laboratory was one realm, and that his home and religion were a completely different one. There are today many vestiges of this nineteenth-century scientific absolutism in our thinking and attitudes. It has given Communism, based on Marx's nineteenth-century background, some of its sense of the inexorable course of history and of "scientific" planning of society.

Towards the end of the nineteenth century, many physical scientists viewed their work as almost complete and needing only some extension and more detailed refinement. But soon after, deep problems began to appear. The world seems relatively unaware of how deep these problems really were, and of the extent to which some of the most fundamental scientific ideas have been overturned by them. Perhaps this unawareness is because science has been vigorous in changing itself and continuing to press on, and has also diverted attention by ever more successes in solving the practical problems of life.

Many of the philosophical and conceptual bases of science have in fact been disturbed and revolutionized. The poignancy of these changes can be grasped only through sampling them. For example, the question whether light consists of small particles shot out by light sources, or wave disturbances originated by them, had been debated for some time by the great figures of science. The question was finally settled in the early nineteenth century by brilliant experimentation which could be thoroughly interpreted by theory. The experiments told scientists of the time that light was unequivocally a wave and not particles. But about 1900, other experiments turned up which showed just as unequivocally that light is a stream of particles rather than waves. Thus physicists were presented with a deeply disturbing paradox. Its solution took several decades, and was only accomplished in the mid-1920's by the development of a new set of ideas known as quantum mechanics.

The trouble was that scientists were thinking in terms of their common everyday experience and that experience encompassed the behavior of large objects, but not yet many atomic phenomena. Examination of light or atoms in detail brings us into a new realm of very small quantities with which we have had no previous experience, and where our intuitions could well be untrustworthy. And now in retrospect, it is not at all surprising that the study of matter on the atomic scale has taught us new things, and that some of these were inconsistent with ideas which previously had seemed so clear.

Physicists today believe that light is neither precisely a wave nor a particle, but both, and we were mistaken in even asking the question, "Is light a particle or is it a wave?" It can display both properties. So can all matter, including baseballs and locomotives. We don't ordinarily observe this duality in large objects because they do not show wave properties prominently. But in principle we believe they are there.

We have come to believe other strange phenomena as well. Suppose an electron is put in a long box where it may travel back and forth. Physical theory now tells us that, under certain conditions, the electron will be sometimes found towards one end of the box and sometimes towards the other, but never in the middle. This statement clashes absurdly with ideas of an electron moving back and forth, and yet most physicists to-day are quite convinced of its validity, and can demonstrate its essential truth in the laboratory.

Another strange aspect of the new quantum mechanics is called the uncertainty principle. This principle shows that if we try to say exactly where a particle (or object) is, we cannot say exactly how fast it is going and in what direction, all at the same time; or, if we determine its velocity, we can never say exactly what its position is. And so, according to this theory, Laplace was wrong from the beginning. If he were alive today, he would probably understand along with other contemporary physicists that it is fundamentally impossible to obtain the information necessary for his precise predictions, even if he were dealing with only one single particle, rather than the entire universe.

The modern laws of science seem, then, to have turned our thinking away from complete determinism and towards a world where chance plays a major role. It is chance on an atomic scale, but there are situations and times when the random change in position of one atom or one electron can materially affect the large-scale affairs of life and in fact our entire society. A striking example involves Queen Victoria who, through one such event on an atomic scale, became a mutant and passed on to certain male descendants in Europe's royal families the trait of hemophilia. Thus one unpredictable event on an atomic scale had its effect on both the Spanish royal family and, through an afflicted czarevitch, on the stability of the Russian throne.

This new view of a world which is not predictable from physical laws was not at all easy for physicists of the older tradition to accept. Even Einstein, one of the architects of quantum mechanics, never completely accepted the indeterminism of chance which it implies. This is the origin of his intuitive response, *"Herr Gott würfelt nicht"*—the Lord God doesn't throw dice! It is interesting to note also that Russian communism, with its roots in nineteenth-century determinism, for a long time took a strong doctrinaire position against the new physics of quantum mechanics.

When scientists pressed on to examine still other realms outside our common experience, further surprises were found. For objects of much higher velocities than we ordinarily experience, relativity shows that very strange things happen. First, objects can never go faster than a certain speed, regardless of how hard they are pushed. Their absolute maximum speed is that of light—186,000 miles per second. Further, when objects are going fast, they become shorter and more massive—they change shape and also weigh more. Even time moves at a different rate; if we send a clock off at a high velocity, it runs slower. This peculiar behavior of time is the origin of the famous cat-kitten conceptual experiment. Take a litter of six kittens and divide them into two groups. Keep three of them on earth, send the other three off in a rocket at a speed nearly as fast as light, and after one year bring them back. The earth kittens will obviously have become cats, but the ones sent into space will have remained kittens. This theory has not been tested with kittens, but it has been checked experimentally with the aging of inanimate objects and seems to be quite correct. Today the vast majority of scientists believe it true. How wrong, oh how wrong were many ideas which physicists felt were so obvious and well-substantiated at the turn of the century!

Scientists have now become a good deal more cautious and modest about extending scientific ideas into realms where they have not yet been thoroughly tested. Of course, an important part of the game of science is in fact the development of general laws that can be extended into new realms. These laws are often remarkably successful in telling us new things or in predicting things which we have not yet directly observed. And yet we must always be aware that such extensions may be wrong, and wrong in very fundamental ways. In spite of all the changes in our views, it is reassuring to note that the laws of nineteenth-century science were not so far wrong in the realm in which they were initially applied—that of ordinary velocities and of objects larger than the point of a pin. In this realm they were essentially right, and we still teach the laws of Newton or of Maxwell, because in their own important sphere they are valid and useful.

We know today that the most sophisticated present scientific theories, including modern quantum mechanics, are still incomplete. We use them because in certain areas they are so amazingly right. Yet they lead us at times into inconsistencies which we do not understand, and where we must recognize that we have missed some crucial idea. We simply admit and accept the paradoxes and hope that sometime in the future they will be resolved by a more complete understanding. In fact, by recognizing these paradoxes clearly and studying them, we can perhaps best understand the limitations in our thinking and correct them.

With this background on the real state of scientific understanding, we come now to the similarity and near identity of science and religion. The goal of science is to discover the order in the universe, and to understand through it the things we sense around us, and even man himself. This order we express as scientific principles or laws, striving to state them in the simplest and yet most inclusive ways. The goal of religion may be stated, I believe, as an understanding (and hence acceptance) of the purpose and meaning of our universe and how we fit into it. Most religions see a unifying and inclusive origin of meaning, and this supreme purposeful force we call God.

Understanding the *order* in the universe and understanding the *purpose* in the universe are not identical, but they are also not very far apart. It is interesting that the Japanese word for physics is butsuri, which translated means simply *the reasons for things*. Thus we readily and inevitably link closely together the nature and purpose of our universe.

What are the aspects of religion and science which often make them seem almost diametrically opposite? Many of them come, I believe, out of differences in language used for historical reasons, and many from quantitative differences which are large enough that unconsciously we assume they are qualitative ones. Let us consider some of these aspects where science and religion may superficially look very different.

Job and Einstein, Men of Faith

The essential role of faith in religion is so well known that it is usually taken as characteristic of religion, and as distinguishing religion from science. But faith is essential to science too, although we do not so generally recognize the basic need and nature of faith in science.

Faith is necessary for the scientist to even get started, and deep faith necessary for him to carry out his tougher tasks. Why? Because he must be personally committed to the belief that there is order in the universe and that the human mind—in fact his own mind—has a good chance of understanding this order. Without this belief, there would be little point in intense effort to try to understand a presumably disorderly or incomprehensible world. Such a world would take us back to the days of superstition, when man thought capricious forces manipulated his universe. In fact, it is just this faith in an orderly universe, understandable to man, which allowed the basic change from an age of superstition to an age of science, and has made possible our scientific progress.

Another aspect of the scientist's faith is the assumption of an objective and unique reality which is shared by everyone. This reality is of course mediated by our senses and there may be differences in individual interpretation of it. However, Berkeley's idea that the world springs entirely from the mind, or the possible existence of two or more valid but discordant views of the world are both quite foreign to scientific thinking. To put it more simply, the scientist assumes, and his experience affirms, that truth exists.

The necessity of faith in science is reminiscent of the description of religious faith attributed to Constantine: "I believe so that I may know." But such faith is now so deeply rooted in the scientist that most of us never even stop to think that it is there at all.

Einstein affords a rather explicit example of faith in order, and many of his contributions come from intuitive devotion to a particularly appealing type of order. One of his famous remarks is inscribed in German in Fine Hall at Princeton: "God is very subtle, but he is not malicious." That is, the world which God has constructed may be very intricate and difficult for us to understand, but it is not arbitrary and illogical. Einstein spent the last half of his life looking for a unity between gravitational and electromagnetic fields. Many physicists feel that he was on the wrong track, and no one yet knows whether he made any substantial progress. But he had faith in a great vision of unity and order, and he worked intensively at it for thirty years or more. Einstein had to have the kind of dogged conviction that could have allowed him to say with Job, "Though he slay me, yet will I trust him."

For lesser scientists on lesser projects, there are frequent occasions when things just don't make sense and making order and understanding out of one's work seems almost hopeless. But still the scientist has faith that there is order to be found, and either he or his colleagues will some day find it.

The Role of Revelation

Another common idea about the difference between science and religion is based on their methods of discovery. Religion's discoveries often come by great revelations. Scientific knowledge, in the popular mind, comes by logical deduction, or by the accumulation of data which is analyzed by established methods in order to draw generalizations called laws. But such a description of scientific discovery is a travesty on the real thing. Most of the important scientific discoveries come about very differently and are much more closely akin to revelation. The term itself is generally not used for scientific discovery, since we are in the habit of reserving revelation for the religious realm. In scientific circles one speaks of intuition, accidental discovery, or says simply that "he had a wonderful idea."

If we compare how great scientific ideas arrive, they look remarkably like religious revelation viewed in a non-mystical way. Think of Moses in the desert, long troubled and wondering about the problem of saving the children of Israel, when suddenly he had a revelation by the burning bush. A similar pattern is seen in many of the revelations of the Old and New Testaments. Think of Gautama the Buddha who traveled and inquired for years in an effort to understand what was good, and then one day sat down quietly under a Bo tree where his ideas were revealed. Similarly, the scientist, after hard work and much emotional and intellectual commitment to a troubling problem, sometimes suddenly sees the answer. Such ideas much more often come during off-moments than while confronting data. A striking and well-known example is the discovery of the benzene ring by Kekulé, who while musing at his fireside was led to the idea by the vision of a snake-like molecule taking its tail in its mouth. We cannot yet describe the human process which leads to the creation of an important and substantially new scientific insight. But it is clear that the great scientific discoveries, the real leaps, do not usually come from the so-called "scientific method," but rather more as did Kekulé's—with perhaps less picturesque imagery, but by revelations which are just as real.

How Much Proof?

Another popular view of the difference between science and religion is based on the notion that religious ideas depend only on faith and revelation while science succeeds in actually proving its points. In this view, proofs give to scientific ideas a certain kind of absolutism and universalism which religious ideas have only in the claims of their proponents. But the actual nature of scientific "proof" is rather different from what this approach so simply assumes.

Mathematical or logical proof involves choice of some set of postulates, which hopefully are consistent with one another and which apply to a situation of interest. In the case of natural science, they are presumed to apply to the world around us. Next, on the basis of agreed-on laws of

logic, which must also be assumed, one can derive or "prove" the consequences of these postulates. How can we be sure the postulates are satisfactory? The mathematician Gödel has shown that in the most generally
used mathematics, it is fundamentally impossible to know whether or not
the set of postulates chosen are even self-consistent. Only by constructing
and using a new set of master postulates can we test the consistency of the
first set. But these in turn may be logically inconsistent without the possibility of our knowing it. Thus we never have a real base from which we
can reason with surety. Gödel doubled our surprises by showing that, in
this same mathematical realm, there are always mathematical truths
which fundamentally cannot be proved by the approach of normal logic.
His important proofs came only about three decades ago, and have profoundly affected our perspective on human logic.

There is another way by which we become convinced that a scientific
idea or postulate is valid. In the natural sciences, we "prove" it by making
some kind of test of the postulate against experience. We devise experiments to test our working hypotheses, and believe those laws or hypotheses are correct which seem to agree with our experience. Such tests can
disprove an hypothesis, or can give us useful confidence in its applicabil-
ity and correctness, but can never give proof in any absolute sense.

Can religious beliefs also be viewed as working hypotheses, tested and
validated by experience? To some this may seem a secular and even an
abhorrent view. In any case, it discards absolutism in religion. But I see
no reason why acceptance of religion on this basis should be objectionable. The validity of religious ideas must be and has been tested and
judged through the ages by societies and by individual experience. Is
there any great need for them to be more absolute than the law of gravity?
The latter is a working hypothesis whose basis and permanency we do not
know. But on our belief in it, as well as on many other complex scientific
hypotheses, we risk our lives daily.

Science usually deals with problems which are so much simpler and
situations which are so much more easily controllable than does religion
that the quantitative differences in directness with which we can test hypothesis generally hides the logical similarities which are there. The controlled experiment on religious ideas is perhaps not possible at all, and we
rely for evidence primarily on human history and personal experience.
But certain aspects of natural science, and the extension of science into
social sciences, have also required similar use of experience and observation in testing hypotheses instead of only easily reproducible experiments.

Suppose now that we were to accept completely the proposition that
science and religion are essentially similar. Where does this leave us and

where does it lead us? Religion can, I believe, profit from the experience of science where the hard facts of nature and the tangibility of evidence have beaten into our thinking some ideas which mankind has often resisted.

So What?

First, we must recognize the tentative nature of knowledge. Our present understanding of science or of religion is likely, if it agrees with experience, to continue to have an important degree of validity just as does the mechanics of Newton. But there may be many deeper things which we do not yet know and which, when discovered, may modify our thinking in very basic ways.

We must also expect paradoxes, and not be surprised nor unduly troubled by them. We know of paradoxes in physics, such as that concerning the nature of light, which have been resolved by deeper understanding. We know of some which are still unresolved. In the realm of religion, we are troubled by the suffering around us and its apparent inconsistency with a God of love. Such paradoxes confronting science do not usually destroy our faith in science. They simply remind us of a limited understanding, and at times provide a key to learning more.

Perhaps there will be in the realm of religion cases of the uncertainty principle, which we now know is such a characteristic phenomenon of physics. If it is fundamentally impossible to determine accurately both the position and velocity of a particle, it should not surprise us if similar limitations occur in other aspects of our experience. This opposition in the precise determination of two quantities is also referred to as complementarity; position and velocity represent complementary aspects of a particle, only one of which can be measured precisely at any one time. Nils Bohr has already suggested that perception of man, or any living organism as a whole, and of his physical constitution represents this kind of complementarity. That is, the precise and close examination of the atomic makeup of man may of necessity blur our view of him as a living and spiritual being. In any case, there seems to be no justification for the dogmatic position taken by some that the remarkable phenomenon of individual human personality can be expressed completely in terms of the presently known laws of behavior of atoms and molecules. Justice and love may be another example of complementarity. A completely loving approach and the simultaneous meting out of exact justice hardly seem consistent. These examples could be only somewhat fuzzy analogies of complementarity as it is known in science, or they may indeed be valid though still poorly defined occurrences of the uncertainty principle. But in any case, we should expect such occurrences and be forewarned by science that there will be fundamental limitations to our knowing everything at once with precision and consistency.

Finally, if science and religion are so broadly similar, and not arbitrarily limited in their domains, they should at some time clearly converge. I believe this confluence is inevitable. For they both represent man's efforts to understand his universe and must ultimately be dealing with the same substance. As we understand more in each realm, the two must grow together. Perhaps by the time this convergence occurs, science will have been through a number of revolutions as striking as those which have occurred in the last century, and taken on a character not readily recognizable by scientists of today. Perhaps our religious understanding will have seen progress and change. But converge they must, and through this should come new strength for both.

In the meantime, each present day, with only tentative understanding and in the face of uncertainty and change, how can we live gloriously and act decisively? It is this problem, I suspect, which has so often tempted man to insist that he has final and ultimate truth locked in some particular phraseology or symbolism, even when the phraseology may mean a hundred different things to a hundred different people. How well we can commit our lives, effort, and devotion to ideas which we recognize in principle as only tentative represents a real test of mind and emotions.

Galileo espoused the cause of Copernicus' theory of the solar system, and at great personal cost because of the Church's opposition. We know today that the question on which Galileo took his stand, the correctness of the idea that the earth rotates around the sun rather than the sun around the earth, is largely an unnecessary question. The two descriptions are equivalent, according to general relativity, although the first is simpler. And yet we honor Galileo for his pioneering courage and determination in deciding what he really thought was right and speaking out. This was important to his own integrity and to the development of the scientific and religious views of the time, out of which has grown our present better understanding of the problems he faced.

The authority of religion seemed more crucial in Galileo's Italy than it usually does today, and science more fresh and simple. We tend to think of ourselves as now more sophisticated, and science and religion as both more complicated so that our position can be less clear cut. Yet if we accept the assumption of either one, that truth exists, surely each of us should undertake the same kind of task as did Galileo, or long before him, Gautama. For ourselves and for mankind, we must use our best wisdom and instincts, the evidence of history and wisdom of the ages, the experience and revelations of our friends, saints, and heroes in order to get as close as possible to truth and meaning. Furthermore, we must be willing to live and act on our conclusions.

What Science Suggests About Us

Science represents an attempt of the human mind to understand and deal with the magnificent, the subtle, the astounding, and yet ordered, universe. The origin and outlook of science has a very strong connection with religious thinking itself. Modern science has in particular grown out of Greek, and then Judeo-Christian cultures. These traditions have no sole claim on science, for there are many other traditions, such as the Muslim, which have contributed importantly. Nevertheless, the Judeo-Christian view that the universe around us is an external reality based on laws which are faithful and constant has probably played a dominant role. This is directly parallel to the concept of an external creation by a creator on whom we can rely and whose laws are constant and inexorable—the creator found in the Bible. But such a view is not universal among the cultures of mankind. And without it, we could easily falter in the faith which still insists on the rationality of nature's laws, and in permanent and reliable principles, when we are faced with apparently unyielding inconsistencies.

By the nineteenth century, mankind seemed to understand these constant laws of the universe well enough, and science had such success in explaining much of nature's behavior based on discovery of apparently immutable laws and logic, that the scientific view tended to become overconfident and rigid. Many thought it only a matter of time for such pursuit of scientific logic to ultimately explain everything. The success of science was indeed remarkable in describing material behavior with ever-increasing precision. This was especially true for what could be then measured most precisely, the motions of individual particles and the behavior of electromagnetic waves. However, the laws being discovered were extending to ever more complex situations and to life itself. A complete and precise view of a rigidly mechanistic universe seemed inevitably to be the end result. For example, the laws of physics appeared to predict with essentially perfect precision just what the course of events would be for a particle of matter once its initial position and motion were known. This led to a hard and fast view that nature's laws required causality, and left in the minds of many no room for any manifestations of a deity. A deity was allowed to do nothing except possibly determine the initial conditions which began the universe on its course. The overwhelming logic of science at that time did seem clearly to indicate a complete determinism, and one which was expected to extend even to details of human life so that human behavior itself might eventually be completely predicted by the rational pursuit of inexorable laws which control us. The faithful laws of the Old Testament as interpreted by science had become overbearing.

By the turn of the twentieth century, peculiar phenomena which fell quite outside the intuition and understanding of nineteenth century scientists began to appear. First, there was the advent of relativity, which concluded that such basic ideas as time and the mass of matter were not so easily defined nor constant. It began to be recognized that not only is the extent of time dependent on one's relative velocity, but also time and distance are somewhat interchangeable. Space itself was no longer so straightforward, but appeared curved, and the finite and infinite became confused. These new concepts, which were disturbing and counter-intuitive, were forced on us by the careful examination of very high velocities or very large forces and distances, something beyond the previous human experience.

A second area where apparently well established concepts fell upon trouble was the realm of the very small, where the new ideas of quantum mechanics were forced on the world of science by experiments which were again outside of previous experience. These new quantum ideas seemed to show that waves such as light or radiowaves always occurred in finite packages like particles. Light which had seemed clearly to be a wave in the late nineteenth century turned out to be made of particles, or at least at times behaved like particles. The most confusing problem of all was that light appeared both particle and wave-like, sometimes manifesting one property and sometimes another. Furthermore, the behavior of atoms and waves, including the radioactivity of nuclei, was shown after all to not be so completely causal in the sense that it was not exactly predictable. What could be predicted was only some probability of a given behavior being followed. Even worse, a basic principle of quantum mechanics states that it is fundamentally impossible to determine the initial conditions of a particle, that is, both its position and its velocity of movement in such a way that its future could *ever* be precisely predicted. This is not a matter of our imperfect techniques nor something which might be improved with time, the indeterminacy is a fundamental law of the universe. Quantum mechanics also says, and very recent experiments have clearly shown, that an event can be influenced by another event so disconnected that we see no possible means for normal causality to have connected the two. Such results were so contrary to usual human experience, previous science, or our intuitive concepts, that they were exceedingly difficult for many scientists to accept.

A paradox often referred to is about Schröedinger's cat—a cat in a closed box and of unknown status. The physicist Schröedinger, one of the originators of modern quantum mechanics, suggested the thought experiment of putting his cat in a closed box with a sealed bottle of cyanide. A mechanism would break the cyanide bottle if a radioactive atom emitted a particle into the mechanism, thus killing the cat. Schröedinger posed such an experiment because the laws of quantum mechanics seem to say that, if we do not open the box, the cat not only has some chance of being alive and some chance of being dead, it is actually in both conditions at the same time. Once we open the box, we will find the cat either alive or dead, one or the other. But that is because we have disturbed the cat's situation—until the box is opened the cat exists in both states at the same time. Can reality actually be this way? Can anything be in two different conditions, alive and dead, at the same time? Such phenomena certainly do not fit any common sense or easily accepted view of what reality should be like. But laboratory experiments say it is real, at least for small inanimate particles or waves. Einstein, who had himself played a prominent part in these revolutions of scientific thought, could never believe that quantum mechanics was actually correct. Nevertheless, the quantum mechanical laws which were discovered, mostly in the early half of this century, predict the behavior of matter and radiation with such precision and surety that any scientific criteria must conclude that they are much more correct than our previous concepts. Can the physical principles governing matter be extended, as in the case of the cat, to living things? Schröedinger asked "Does the cat think it is alive or dead." We now understand that for anything as large as a cat, this anomaly can occur for such a short time that we would never sense it. But the queer conclusions from quantum mechanics can at times clearly determine the behavior of large-scale human events, and we can be sure that at times they have.

Thus, both in the case of relativity where experiments were extended to the realm of the very large or of very high velocities, and in the case of quantum mechanics where our experience was extended to the realm of the very small, it turned out that our whole understanding of scientific laws was to change. These drastic changes, an almost complete revolution in the philosophical outlook of science, are now almost universally accepted because of the overwhelming experimental evidence and the success of these new and revised laws. At the same time, older forms of scientific laws which had previously been verified by experiment and relied on up until that time were quite correct in their own field of application. This was a replacement of the old laws by new laws which were more general. It left the old laws as a good and valid approximation where they had previously been used, but quite inadequate in new and important realms which science later began to explore. And we know of remaining inconsistencies which show us that our present understanding is still incomplete.

The uncertainty principle, stating that initial conditions simply cannot be defined sufficiently well to make final or accurate predictions for the future, was seized on by those who wanted a place for divine intervention. Since the future could be predicted only in terms of a probability for following various courses, it was tempting to believe that there was an opportunity for the influence of a deity. Was the hemophilia of Victoria'a male descendants, for example, a case of divine intervention which changed the course of history? It was dependent, apparently, on a small event on an atomic scale. But in fact logic did not really seem to allow a place for divine intervention anymore than before. It simply showed us that the course of events is unpredictable and that our previous thinking and conclusions, overwhelmingly correct though they seemed to have been at the time, were much too limited. Laws and events which were previously beyond our imagination turned out to be real. No sensible scientist of the nineteenth century could possible have accepted many of the things we are now convinced are true; some of our present experiments and conclusions are simply outside the past realm of serious thought. How things can be in two states at the same time—alive and dead— is still a troubling question.

It is fantastic that the human mind can understand so much of the logic of our universe, and force itself into new modes of thought as a result of inexorable logic coupled with careful observation. It is also remarkable that we can from scientific evidence trace the history of our universe back far beyond the time when humans or even the very first signs of life occurred, back almost to the very beginning. This tracing is by no means a hypothetical construct, it is based on specific observations and their logical consequences. And recent observations have revealed a unique event in the history of our universe which has further changed the outlook of the scientific community. One set of experimental observations has shown us that other galaxies of the universe are essentially all moving away from our own, and hence the universe must be expanding. Measurement of the rate at which they are moving away from us and their distance suggests that everything must have been very close together, perhaps at a single point, only about 15 billion years ago. This was again contrary to the expectations of many scientists, and hence a difficult conclusion. It called for some kind of unique event 15 billion years ago when the universe might be said to have "started." Many physicists and astronomers looked hard for alternate explanations to show how an always constant universe might only *appear* to be expanding. Various theories to that effect were constructed and subjected to tests. However, somewhat by accident, what seems to be very clear evidence that there was such an expansion and a unique time in the past for our universe was recently discovered. This was the detection of radiation left over from the time of this momentous initial explosion, the "big bang," which started all matter on its expanding way, gradually forming the galaxies and stars as it expanded. Discovery of remnants of this unique event in the past, which might be considered the creation of our universe, again raised the possibility of the role of a deity, at least in initiating such an event. However, scientists are busy trying to penetrate even behind this cataclysm; we do not yet understand whether such efforts can succeed, but some further evidence of conditions at the time is accumulating.

How far can science go? Will it be able to see behind the "big bang" and give a satisfying explanation even for the existence of our universe? How far can it go in understanding life? We know in increasing depth many of the life processes, and find that much of life can be reduced to intricate chemical reactions. Yet the very formation of life is still unknown. Perhaps also, there are new phenomena and principles still quite unimagined in the overall organization of a living being. Perhaps here again, our basic ideas and philosophical outlook will be radically changed as we explore further. Is it fundamentally possible for an object, such as the human mind, to really understand itself? Will science eventually give us a satisfyingly complete picture of all processes in the universe or, as we explore further and understand more, will the still mysterious bounderies of our knowledge only continue to expand?

As our understanding of the universe has steadily improved, experience has taught scientists to be somewhat more humble about how thoroughly they really understand. We have seen what appeared to be unshakable principles completely changed. We know that even our present best theories have apparent inconsistencies. Yet in this change, the old laws of physics, for example Newton's laws of motion, still apply very well in the places where they were previously used, and they continue to be honored and applied. They were simply workable approximations coming out of a somewhat limited view of the whole picture—we knew in part, and we still know only in part. The radical changes in our outlook and understanding of nature have given us the ability to understand areas of the universe which were previously quite unknown and unexplored. At the same time it has shown us that human knowledge must always be tentative, though it is also very powerful. The extent and precision with which science now relates and predicts the behavior of many aspects of our universe by a relatively few simple laws is enormously impressive. And while much of the excitement of science is in exploring new ideas, such new ideas are examined with critical thoroughness and in the light of a wealth of experimental evidence since we are very conscious of the strength and the evidence for present ideas.

Analogies to religious thinking are perhaps readily seen. First, while we may understand much and we can have faith in the past experience of many human generations and well-tried principles, yet we must regard our understanding as tentative and possibly subject to drastic reorientation. Scientific knowledge is no longer so absolute and monolithic. I believe we can assume our human religious views are similar; "only God is ultimate." Furthermore, if we are to understand more we must always be open to new ideas and new experience. But if so, then as in the case of science, we must expect our present understanding, or religious laws if you like, both to have genuine validity in those areas where they have been tried and tested, and yet at the same time to be of limited scope, truth seen through a glass darkly. We must expect troublesome reorientations of our thinking and discovery of new depths of understanding in areas which have not previously been adequately understood. As in the scientific realms, we may be called to have faith and to act carefully and thoughtfully in the light of present understanding, yet to be always open to radically new insights. There may be some who believe religion has been revealed to us in a way which needs no further change. Even if this is the case, the understanding of an adult must be different from that of a child, and many individuals have become confused by the transition from the understanding of childhood to that of an adult. To what extent can our religious understanding undergo non-disruptive revolution as has science, where radical and exciting change has occurred without a loss of either faith or effectiveness? I believe that is a critical question in modern times. In the religious realm, and in the faith of individuals, how might there be exciting, revolutionary, yet non-disruptive change where the continuity of sustaining faith and faithful behavior is maintained?

And where does our scientific experience now leave us? As for myself,

I thoroughly believe in the penetrating validity, yet tentative nature of science. I also believe in the possibility and importance of a close and personal relation with God. Is this combination naïve? Perhaps, but I think not as naïve as denying such a possibility. The weight of evidence, while certainly not enough to convince all thoughtful people, is, I believe, in that direction. Am I susceptible also, one might ask, to a penchant for astrology? No. By contrast with our rather limited understanding of the creation and of the nature of the human spirit, we understand in great detail the nature and motions of planets; evidence tells us very precisely the forces they seem to exert, and there is simply no reasonable room for them to have the effects assumed by astrologers. The validity of God's relation to us and the non-validity of astrology are both human judgments which must be based on experience and carefully made. They are not absolute, but to me they are both strong conclusions.

Consider now the more material effects which have come from science, what it has created rather than how it has affected our philosophical outlook. Increasing knowledge has fantastically enlarged mankind's potential. It has produced remarkable improvements in man's health. This now even includes ways of eliminating genetic handicaps and defects. Scientific and technical developments have made it possible for us to produce food for billions of people on earth—for everyone presently living and many more. We have expanded man's power and control over the natural elements. A single individual can, with the help of machinery and technology, accomplish more work than whole armies of slaves could previously have done. We have increasing power to create new forms of life. Even the potentiality of the mind has been much increased, for example by the advent of computers which take care of many of the routine processes with which we were formerly burdened, and carry out processes of which we have never before been capable. This opens up new possibilities for the better uses of man's energies and creativity. We can now travel with ease and speed to any part of the globe, communicate cheaply and quickly anywhere, including beyond the solar system, if creatures are there to receive our communications. The range and power of essentially all our activities have been increased. Mankind has conquered the air by flight, space with rockets, landed on the moon, and will probably spread the human habitat to other planets. We probably cannot foresee the many major scientific or technical advances which are coming next, but must expect many more near-miraculous developments. In most directions, what we might accomplish appears not to be bounded by limitations of science and technology, but rather to depend primarily on mankind's own interests, ideals, and drive—these are the things which will typically shape our actual course. In a very real sense man, who has been created in the image of God, is ever more a co-creator with God.

Knowledge of science and technology, mankind's brainchildren, has become a primary force and a primary wealth of the nations which will shape the future. As Shakespeare's Hamlet exclaimed "What a piece of work is man! How noble in reason! How infinite in faculty...in apprehension, how like a god!...and yet" "...and yet" Hamlet says, and begins to deal with his uncertainties.

Our increased power and potential can enormously enrich life, *and yet* threaten to despoil the earth of its beauty and resources. Our increased medical skill prolongs healthy life and and our ability to produce can feed a large population, *and yet* long life can be a misery and the present population is still primarily limited by poverty, starvation, and war. Plain, ugly, ordinary war has destroyed tens of millions of people during the last several decades. But the threat of massive destruction by our discovery of how to tap nuclear energy is still more frightening; chemical and biological warfare would perhaps be equally so if the public attention were focused on them. Einstein has even compared the effects of advances in technology with putting an axe in the hands of a pathological idiot. And so, we are faced ever more urgently with the problem of understanding the purpose of our universe as well as how its parts operate and interact. Knowledge presents us with inspiring possibilities as co-creators with God. As a result, devotion to inspiring purposes and a stronger grasp of religious wisdom are ever more pressing. And "where is the wisdom we have lost in knowledge?" How important it is for those of us who are scientifically or technically skilled to understand and participate in God's purposes, and how important it is for those whose special role it is to study religion to also understand the context, the problems, and the challenges of mankind's creative scientific future! We might guess, but we don't really know what miracles and problems will occur in the future. But any reasonable evaluation of the rapid increase of science and technology during past decades must lead us to expect further remarkable developments. Should there be increased attention in training of religious leaders to this part of our nature, and increased contact with scientists so that their religious understanding is adequately developed?

It is frequently said that science itself is amoral. Let me present a rather different point of view. There are aspects of science which, if they do not represent ethics itself, are at least very close to ethical dimensions. Scientific examination of living has replaced some of the more formal rules of religion. The dietary laws of Moses are now better understood; scientific knowledge of health and diet have at least partly replaced them and given us important new insight. Living in peace, and in peace with oneself, is also subject to careful thought, observations, and analysis. But still more interesting from the point of view of ethics, I believe, is the discipline of truth. Reason, particularly reason coupled with experimental tests and observations, hold scientists to very high standards of truth and objectivity. They provide real punishment for either deception of others, or for self delusion and catering to ones' own particular wishes about what might be true. The progress of science also responds strongly to openness and sharing. To what extent is the discipline of truth equivalent to ethical behavior? It enforces values which are demanding, not self-centered, and generally beneficial to the human community. It is perhaps only the poet who can baldly say "Beauty is truth, and truth beauty." However, many scientists feel that.

While science in general is quite ancient, modern science, with a clear concept of experimental observation, is only a few hundred years old. Its progress during these several hundred years has been spectacular. Experimental tests have provided a discipline for scientists and the possibility of constructive addition to the work of individual scientists into an edifice which has been cumulative. It is this human and social context, including the sharing and the intercomparison of ideas, which has been the foundation of success. It has thereby grown into an impressive structure of knowledge and understanding, provided enormous power and flexibility to mankind, and with its helpmeet technology now plays an almost dominant role in our civilization. The effect of this discipline of truth on the attitude and thinking of individual scientists has also been powerful. There are of course a few unethical scientists, unethical in the sense that they are willing to depart from the truth and either overtly or carelessly depart from objectivity. Some frauds in science have recently been attracting attention of the popular press. They do occur, but they are relatively rare and are self-corrected by the overall attitude of the scientific community which tends to be quick and relentless in pointing out mistakes or frauds and thus maintaining the ethics and progress of the scientific community. I want to emphasize that ethical scientists frequently make honest mistakes in the course of work. Such mistakes are corrected by colleagues, or often the scientist who makes a mistake may be the quickest to recognize and publicly correct it. One of my colleagues announced an important discovery of ammonia on the planet Mars, which would indicate the exciting presence of life there. But he was also the quickest to publicly announce that he was mistaken; he later discovered he had misinterpreted his carefully taken data. In some fields, mistakes are frequent and when the limitations in ones grasp of the truth are clearly and openly stated by the scientists himself, they are not unethical nor damaging to the structure of science as a whole. Rather, the willingness to venture surmises and recognize them as such is important, and so work which might turn out to be in error can be a part of the process of rapid discovery when its limitations and uncertainties are carefully noted. On the other hand, any overt falsification or unwillingness to recognize error when it occurs is very severely treated by the scientific community. The individual is not severely treated as a person, but other scientists recognize that his work can be ignored, and his career as a scientist is not valued or encouraged.

I would not pretend that scientists in themselves are more ethical than those in any other field. We are simply fortunate that experiment provides well-accepted tests of truth. In scientific work itself one does not usually dare to be unethical in the sense of avoidance of the truth because correction is generally sure and swift. This in itself develops an atmosphere of devotion to the truth, not overtly associated with the thought of punishment, but because of its accepted value. Of course, as I indicated earlier, we recognize that scientific truth must be regarded as provisional. Furthermore, there are undoubtedly errors in our present results and thinking. Nevertheless, I believe it is precisely the existence of broadly accepted tests for truth and a community which has developed the habit of being self-critical that allows us to eventually arrive at broad agreement, and that has provided the remarkable growth and success of science. Is there any analogy in the domain we recognize as religious?

The Bible has many stories about tests of the true God, or of religious truth. There is the rod of Aaron which became a serpent and swallowed those of the Egyptian magicians. There is Elijah and the test of his sacrifice against that of the priests of Baal—a test so overwhelming that it resulted in slaughter of those promoting false religion. There are the miracles of the New Testament. But we seem to have no recent events which provide such clear demonstrations. Any tests of the validity of religious views are characteristically much more complex and arguable than is the case for the physical sciences. Perhaps frustration over the difficulty of clear and simple demonstration of religious truth is part of the reason for many biblical accounts of convincing miracles. There are, however, scientific fields with somewhat similar difficulties—where large amounts of observations accumulated over a long time and assembled in a convincing pattern is the only evidence available—for example, this was the characteristic of Darwin's case for evolution. I have always regarded the Bible as just such a wealth of illustrative cases of very human behavior and its consequences. Human experiences with living or behavior allow some conclusions as to their results, but unfortunately this may be evident only after an entire lifetime or possibly, as Exodus puts it, unto the third or fourth generation. Yet human experience and history, in addition to Biblical illustrations, do provide us with a basis for judgment, somewhat like experiments in the case of the simpler sciences.

Even among literalist Christians, there is no universally agreed interpretation of some parts of the Bible. And I do not myself find very interesting or edifying most of the classical intersectarian disputes about doctrine. There are many arguable aspects of religious views. Nevertheless, there is also clear and general consensus about important moral principles, for example about truthfulness and untruthfulness, or about selfish uses of either oneself or of others. And there *are* behavioral tests—for example Christ's suggestion that "by their fruits you shall know them."

How much real analogy is there in the religious field to self correction and the clarification of scientific truth by the community, which has been critically important to the spectacularly successful growth of science? In the case of the religious community, clearly the quality of individuals selected for theological or ministerial training is critical. The quality of individuals attracted to science is just as crucial. However, it is the purifying effects of commonly recognized standards, and the clarity with which success or failure depends on maintaining these, on which science has been built. To what extent can care about common standards and intercomparison of ideas provide similar strength to the religious community? Can these behavioral standards be made infectious in theological schools, or do they come only from a still broader effort so that we hand over responsibility to the family, the churches, or even the lay community?

The scandalous and hypocritical action of some television evangelists recently have illustrated forcefully the destructive effects of lack of standards. They have made the word "T.V. evangelist" somewhat repulsive, and discolored the word "evangelist" in the popular mind. Such cases are clear enough, and the disasters which have resulted will presumable help deter such hypocritical and inconsistent behavior in the future. To what extent are we sensitive in areas which are not quite so blatant? How should the Christian community, theological students, or professionally trained theologians and ministers react to one of their members who, in his or her enthusiasm is tempted to try to support what is perhaps a very worthy cause with statements that are only marginally true or not true at all? False statements are unfortunately all too common in the political world, and when Christian forces operate in the political or near-political realm as they must on occasion, it is all too easy to disregard truthfulness while selling a desired conclusion. I have heard explanations given for such cases that the cause was right and important, so hence a few misstatements don't really make any difference, or that there just wasn't time to check on the accuracy of information which seemed in the right direction and persuasive. But I don't believe the commandment "Thou shalt not bear false witness..." has been modified in Christian tradition by "unless you feel it's useful to your cause or unless you don't happen to be sure." To what extent should the Christian community react and correct misstatements, perhaps done for just causes? To what extent does it explain them away as an inevitable part of the prophetic tradition? To what extent does this in the long run undermine confidence in religious leadership which our communities and the world need so badly? To what extent does carelessness with truth even prevent the individual from arriving at genuinely Christian conclusions? Are we clear about how we do or should be observing other Christian standards or ideas? Consider, for example, a minister who has twice broken sacred marriage promises with two divorces and three marriages. Is this the leadership we look for in human relations? Is it what we should look for in broader questions such as international peace, where also the keeping of agreements and treaties are important, as is the understanding that selfish interest of an individual nation must be subservient to the good of all nations. How should we be oriented to react in such situations so that both the individual and the Christian community can fulfill the roles to which they are called?

It is easy to recognize the severity of the burden which falls on individual ministers expected by their congregation to be shining examples of Christian behavior, and at the same time partisan to every special interest and foible of the congregation. And of course they as individuals can be expected to make mistakes. The strength of science, as I see it, is not the perfection of an individual scientist, but rather that scientific mistakes are usually honestly recognized, examined, and rectified—first by habits of the individual which spring from acceptance of standards and the context in which he lives, and if not, then perforce by the community.

I do not pretend to know how much further beyond the present earnest efforts religious training and the religious community can reasonably go in this direction and by it towards providing a community of increasing strength and inspiration for us all. Rigid orthodoxy enforced by religious zeal is certainly not what is needed. But I do know the power of self-correction in the scientific community, and of continual thoughtful tests of scientific standards. I also believe that in all fields and for all groups, we are eventually known by our fruits, which in a sense represents an experimental test. And the effects of individual scientists and of science as a whole, of individual religious leaders and of the various religious approaches among us, will all inevitably be subject to such a test.

Why Are We Here?
Where Are We Going?

"Why Are We Here?" and "Where Are We Going?" are two awe-inspiring questions which no discourse can ever really cover. The practical question is whether something can be said about the subject which is clarifying and useful.

Consider first the past history of our universe. Science has given us remarkable knowledge of the past and at the same time helped identify where we are ignorant about it. With substantial evidence and reasonably comfortable surety, we can trace the history of our universe all the way back to what might be called the origin. At the same time, there are prominent phenomena which are not explained in terms of what we presently understand. Perhaps still more importantly, we have essentially no knowledge about why the universe has the precise properties which it has, and yet these properties are critically important to our existence itself.

We detect, coming from all directions, a spectrum of relatively long-wavelength radiation from the so-called big bang—a possibly unique period about 15 billion years ago when our universe was extremely small and violently exploding. As the universe expanded, globs of material gathered together to form the great galaxies we see today. We can even look into the far distance with our telescopes and see galaxies as they were about 80 percent of the way back in time to this very beginning. That such a unique explosion was real is supported not only by the remnant radiation we detect, but also by the relative abundances of the hydrogen, deuterium, and helium which we see today. These abundances correspond remarkably closely to theoretical predictions for such an explosion. Nevertheless in general the scientific community is instinctively opposed to believing that there was ever any grossly unique period or situation in the universe. That seems too arbitrary and improbable. Such a feeling has led to considerable effort to avoid the peculiarity of a unique big bang and formation period. Since evidence seems to force the conclusion that there was indeed an explosion, one possible way of avoiding a unique time is to suppose that the universe, while expanding at the moment due to this explosion, will eventually come to rest because of the attractive forces of gravity and then begin to contract and fall in upon itself to make again a very small object. This in turn will bounce back in an explosive way and initiate another expansive stage, so that our own period is not in fact unique, but only one of many such cycles of the universe. We don't at present know any mechanism that would make it bounce. We also do not find enough mass in the universe to adequately slow down its expansion and subsequently produce the postulated contraction. What we directly see is only about one percent of the required mass, though we have observations of motions of galaxies which indicate there is in fact another ten times more mass which we are not seeing. Thus, we have good evidence for as much as perhaps ten or even 20 percent of the mass required to produce the repeated expansions and contractions which some scientists postulate. If this contraction is really to occur, there must be either an additional phenomenon we have not allowed for or additional mass which is not being detected, perhaps of some new type about which we are still fundamentally unaware, and for which scientists are searching actively. In the meantime, we remain uncertain.

We do in any case understand much of the development of the universe from the big bang to the present time. As it expanded, some of the materials contracted into large structures the size of our galaxies, with masses as great as 100 billions of stars. What produced this clumping of material into the galaxies we see is another one of the intriguing problems not presently understood. However, after such clumping occurred then we believe we understand the main subsequent events. The gaseous material contracted further into stars due to gravity. Within the interior of stars, conditions are such that nuclear reactions take place and make them very hot. From the original hydrogen and helium of the big bang, these reactions manufactured the heavier elements such as carbon, oxygen, and nitrogen. In time, some of these stars became unstable, emitting gases or exploding as supernovae, and thus throwing enormous quantities of the newly made materials into interstellar space, and producing clouds containing not only hydrogen and helium but large amounts of the many chemical elements we now know, including in particular carbon, oxygen, nitrogen, iron, and others which are so important to our own bodily structure and functions. Such clouds then contracted again into second-generation stars of which our sun is one, and around the stars were further condensations of materials such as the planets circulating around our sun, also containing the newly manufactured atoms of carbon, oxygen, etc. Thus, the nature of our earth depends on our sun being a second-generation star formed from raw materials first manufactured by earlier stars and thrown out into interstellar space by them.

That our universe developed the way it did is delicately dependent on certain particular characteristics. For example, during the first second of the initial explosion, the explosive energy and the amount of mass had to be very closely matched, to better than one part in a trillion. If the mass had been slightly too much the universe would have almost immediately contracted again; if it had been slightly too little, the universe would have expanded so rapidly that there would have been no condensation of matter and no formation of stars or life. The electromagnetic forces and the nuclear forces also need to be very delicately balanced. If nuclear forces were slightly too weak, the heavier nuclei would never have formed and the universe would have been made only of hydrogen. Chemistry would have been remarkably simple and uninteresting, with essentially only one element. On the other hand, if the nuclear forces were a bit too strong then we would have had only very heavy nuclei and hydrogen would not have existed at all. Similarly, the strength of gravity had to be just right. Our own sun produces energy in a way which depends on a certain balance between the force of gravity and the rate of generation of energy by nuclear reactions in its interior. If the force of gravity had been somewhat larger, then stars would have cooked the nuclear fuel much more rapidly and their lifetimes would have been unhappily short. Our own sun, fortunately, has already lasted 5 billion years—the time needed for life to develop on one of its planets—and we can expect about another 5 billion years. If the force of gravity had been weaker, material would probably not have clumped so nicely. Or, if galaxies and stars had formed at all in this case, they would have been of much larger mass and processes would have been much slower, so that we could not have developed to the present stage. A large number of other details in the laws of physics fortunately turn out to be just right. Overall, the properties of our universe need to be strikingly and delicately adjusted in order for life anything like we know to exist. Such an observation has been the basis for what is often called the anthropic principle—that is, the laws of the universe are arranged precisely such that humans can exist. Of course, one can also look at this as simply a tautology: since we exist the universe must have laws which allow us to exist. But what is striking is that the laws and parameters of our universe are so precisely those which can produce life. Again,

such a sense of uniqueness is against the instincts of most scientists, because it seems so highly improbable. This is one of the reasons that the
existence of many universes have been postulated—each perhaps somewhat different. We are only observing one of them, namely the one in
which human life in our form can exist. The others are present in multitu-
dinous forms and exhibit other phenomena, but not the formation of life
which creatures like ourselves can experience.

Our own earth is remarkably full of life. Is there life on other planets
around our sun or on planets about other stars? This has been a persistent
human question. We now know that the clouds of interstellar material
spewed out by past stellar explosions are rich in molecules of the type
which we believe are required for the formation of life. In interstellar
space we see carbon, silicates, and other materials which condense into
dust somewhat like that under our feet. In addition there is a wealth of
gaseous molecules, including water, ammonia, and many hydrocarbons
such as alcohols, ethers, and powerful substances like hydrogen cyanide
and formaldehyde—many of the common and active chemical materials
from which biologists believe life was originally formed. Hence, it is easy
to surmise that, as this material condenses into second generation star and
planets about them, the planets are rich in a wealth of molecules and organic materials required for the building of living beings. Furthermore,
we can see on our own earth the fantastic adaptability of life forms, some
living in the coldest parts of the earth, some in the hottest parts, some surviving by using oxygen as we do, others surviving with very little oxygen
or by using carbon dioxide, some at high pressure under the sea, some at
high altitudes in rarified atmospheres, some in darkness, and some in
light. The adaptability of life is indeed impressive. However, the wide
ranges of conditions we find on earth are by no means the most extreme
conditions on planets in general. By now, space exploration has given us
extensive enough information about other planets in our solar system to
conclude that, in contrast to our earth, the present existence of life on
these planets or their moons is quite unlikely.

We believe that we understand how planets are formed and how some
of the smaller ones such as our earth would have lost most of their former
hydrogen and helium and become firm, earth-like objects surrounded by
an atmosphere. There seems every reason for some stars other than our
sun to have planets around them which are favorable to life. And we understand how life can be supported on our planet or possibly others under
a wide range of conditions. But we must differentiate between the possibility of the *existence* of life, which perhaps we understand, and the possibility of its *initiation*, which we clearly do not understand. We simply do

not know the detailed events which initiated the formation of life—that is, a complex molecular organism which can reproduce itself and evolve into something like the earliest fossil forms we know. We generally understand the nature of molecular reactions and can imagine what might take place to build up the more complex molecules necessary for life. But for the present, just how this actually occurred and the probability all are quite beyond us. Sometime ago, Lecomte du Nouy tried to point out that the molecular structure of life is so complex and the probability of each item being properly put into place is so exceedingly small that some completely mysterious creative event had to take place. However, we know that crystals form very easily in environments we understand well, and they too have many atoms which have found their way into exact positions. Crystal regularity and growth depends in an understandable way on thermodynamics; no mysterious event is needed. For molecules important to life, there may also be natural mechanisms which push the atoms into proper place so that the probability of formation is not necessarily extremely low. But for these molecules of life, we do not know the critical steps and cannot at this point make any reasonable calculation of the probabilities involved.

There has been much speculation about the possibility of life elsewhere in the universe, and a general feeling that somehow life must not be uncommon. We know that there are an enormous number of stars more or less like our sun. Our own galaxy contains as many as about 100 billion stars and there are perhaps as many as 100 billion galaxies. We understand mechanisms whereby planets can form about single stars, though in fact we have never detected any planet of the size of our earth around any star like our sun. Detection of such planets is very difficult and we have little clear evidence of the existence of any planets other than those around our sun. Nevertheless, we can reasonably believe many must be present in our universe. The probability of life existing about other stars is hence primarily determined by the probability of its formation. This is the key question and still unanswered.

We do have three important pieces of evidence about the chances of life forms existing on other nearby stars or perhaps anywhere else in our universe. The first piece of evidence is that we ourselves exist. We do know that complex life forms appeared on our own planet about $3\frac{1}{2}$ billions years ago, only about 1 billion years after the formation of the planet itself. They have been evolving ever since in the long, slow, and yet remarkable process of evolution which has developed the many forms we see today, including our friends. So life can form, and again scientific instincts abhor the uniqueness which would be involved if it is only we who exist and there is nothing else similar anywhere.

The second pertinent piece of information is that all life on earth is related, and must come from essentially the same origin. This we know because many of the critical molecules involved in living organisms have a certain symmetry which we call left-handed; none of them are right-handed. Yet, it's clear that the right can be just as effective as the left hand and there is no reason why life should not exist in a righthanded form. So, since all life we see from the simplest to the most complex forms involves left-handed molecules we must all be related and have come from the same event of origin. This tells us that the probability of formation of life, even on this earth where conditions for life seem to be very favorable, is extremely small. It cannot have been initiated more than just a few times during the earth's $4\frac{1}{2}$ billion years. Otherwise surely both types would have survived and would see both left-handed and right-handed types. Hence the probability of formation must be small and probably requires special conditions. But we do not know how small nor do we know much about those conditions accepting that life was formed at a time when there was little free oxygen on earth and probably conditions were quite different from those we presently experience.

The third piece of evidence is somewhat less direct but may shed some light on the question of the existence of other forms of life much like our own. It is simply that no one has yet contacted or visited us from a distant planet. If life in other places were like our own, it would be curious. And once it develops the type of scientific knowledge we presently have, further knowledge should grow rapidly. Our own knowledge and technical abilities have multiplied enormously in the past hundred years; consider what they may be in a million, or even a few hundred years. Any planet which is only one million years ahead of us in this respect would presumably have a form of life vastly superior to ours from a technological and scientific point of view. In our own lifetime we have found a way of reaching the Moon; we should hence expect other life like ours but only a million years older may be visiting other planets, including those on other stars. And those extraterrestrials should be curious to visit even us. Yet we see no one.

Thus, there's considerable logic to the idea that the probability of lifeforms like ours is very low and life itself is rare enough that we have a special role. But even though rare, there may be a number of colonies of life on a few of the billions of stars right in our own galaxy, still so far away from us that perhaps their migration and expansion has not yet reached our own locality. We are in the process of listening for any signals which might have been put out by life on other planets. So far, we find none.

Let us now try to summarize the present stage of us humans. While we understand a great deal and our understanding is growing rapidly, there are still many basic questions which are open. How were galaxies ever formed? Is there missing mass or will the universe continue to expand, gradually cool and eventually die so far as life is concerned? Why do the physical laws have the characteristics they do? What other basic particles or forms of matter have we still not seen or even imagined? How are quantum mechanics and gravity properly connected? We find no consistent way of bringing these two important fields together. How did life begin? Even in the absence of such fundamental knowledge, our remarkably detailed and thorough understanding of many physical processes allow us both to make some very accurate predictions of the future and at the same time to recognize a basic lack of predictability in our universe. This lack of predictability comes about partly because of the uncertainty principle of quantum mechanics, and partly because of a characteristic of complex phenomena physicists call chaos. The latter can be characterized by our attempt to predict the weather. At present, weather can be reasonably predicted over a period of a few days, not very well predicted over more than a few weeks, and for still longer periods not predicted much more accurately than by our broad knowledge of seasonal variations. This is because, as is the case for many aspects of life, very minor effects can produce major results. It is sometimes said that the flap of a butterfly's wing in China may be a critical event in eventually producing a major storm in the Atlantic. The effect is very much like the traditional horseshoe nail, where for lack of a nail the shoe was lost, for lack of the shoe the horse was lost, for lack of the horse the rider was lost, for lack of the rider the battle was lost. Hence, the war was lost; hence, the kingdom was lost, etc. All for the lack of a horseshoe nail. And who could have predicted such a path of events or the many other more complicated and sometimes unstable paths involved in human circumstances?

There are other complexities which we do not understand well enough to even discuss clearly. Some may be imbedded in the nature of free will. No scientist feels that he can justify the common idea of free will on the basis of known principles. Yet I believe almost every scientist, perhaps somewhat unconsciously, acts as if free will occurs. We intuitively accept the idea while recognizing that in our present framework of understanding it cannot be correct. Are basic new concepts involved? Do we simply fool ourselves? What really is free will?

In the description of the majestic universe which was briefly outlined, we believe that we understand rather well some of what is happening and can describe many aspects in detail. Yet the simplest forms of life are enormously more complex than any of these matters which we can successfully describe at present. No doubt in time we will understand much more, and we can anticipate that with excitement. Are there new phenomena imbedded in complexity which we have not yet grasped, just as we did not dream of quantum mechanics before we could examine the very small? Our brains contain about 10^{14} synapses or, in modern computer jargon, 10 million megabytes of information—and it is all interacting. Can the human mind understand with any completeness something as complex as itself, or is complete understanding of any system by itself fundamentally impossible?

Our scientific and technical success has opened up wonderful new possibilities. Some we already enjoy and some we can anticipate. On the practical side, it has also brought new problems and the realization that still further problems may develop. There is population growth, coming out of our successes but producing serious problems. There is environmental damage, which can injure both the environment and ourselves. There's the possibility of self-destruction by the enormous power of nuclear weapons or biological and chemical poisons. There are also major natural threats to humanity which are still not subject to our control. The outburst of AIDS provides a thought-provoking example of a potent virus with no ready cure. Fortunately, that virus is difficult to transmit and requires special circumstances for infection. But imagine a new virus, equivalently dangerous, which can be transmitted simply through the air. We presently have little way of predicting the possibility and potency of such attacks. Thus, the human race is enormously successful, its biomass on earth represents in some sense a fantastic evolutionary success. Our understanding results from impressively successful deductions and potent knowledge. Nevertheless, we are still faced with large uncertainties and great dangers, some which are inherent in our successes.

How far and to where does our universe allow us to proceed? I believe it clear that we will grow further in knowledge, particularly in scientific and technical capabilities. We will continue to increase our potency for doing things, and the variety of things we can do. It seems very likely that the growth of such knowledge cannot be stopped. It's too widespread and interactive with all that we are. Furthermore, our scientific and technical development can be expected to be rapid and probably even increasing in speed. A perspective is provided by remembering that all of written history has occurred in only about 70 human lifetimes. Most of science has developed within the last couple of hundred years, and a large fraction of it during the lifetimes of those of us here. More scientists are said to be alive today than have ever lived before. The growth of our knowledge has tended to be exponential. Furthermore, without some catastrophe the human race has many years to go. Our sun is already about 5 billion years old but it has another few billion years of much the same benevolent support for us. By the time it begins to cool down we will perhaps already be in collision with the Andromeda nebula and there will be still other problems. But that is a long way off and perhaps we will find ways of solving all of these distant problems.

The future possibilities are certainly impressive. For example, if we learn how to produce fusion energy well, then only a few swimming pools full of water can support all the power we now use on earth for a thousand years. We have recently learned how to control the very small. That is, we can manipulate individual atoms and molecules, placing them in particular configurations. We examine and measure down to a hundredth the size of an individual atom or less. Very likely we can also learn to control things on a very large scale. While the planets Venus and Mars are not presently suitable for human life, we can speculate that the science called terraforming may make such planets more like the earth by modifying their atmospheres and surfaces in a way that will allow suitable colonization by humans. Just how far we can go with such planets or moons is still not known, but it is not unreasonable to believe that we can transform some of them into quite livable real estate. We may migrate to planets of other stars. Here on earth, we are also increasingly able to modify the forms of life. While only at its beginning, our understanding of biology and biotechnology has revolutionary potentialities.

The scientific puzzles I have noted above as well as many others will likely be solved, and their solutions allow us to discover still deeper puzzles. It is not clear whether we will ever reach the end of scientific understanding. I personally think and hope not because further exploration of knowledge is so fascinating, motivating, and enjoyable. But there will also be bounds to our potential knowledge and to control of our future as we have found in part from our understanding of quantum mechanics. Our understanding of complexity should advance remarkably, helped especially by the development of computer systems and the research directions they have stimulated. To what extent will humans succeed in understanding with any completeness our own minds and personalities? Can a complex system ever have enough power to understand itself? We can be sure of continued and remarkable advances in human understanding, but we can also expect that deep mysteries may remain.

Increases in knowledge with consequent increase in human abilities to manipulate both our universe and the nature of human life itself presents us with increasingly big decisions which will necessarily be based partly on knowledge, but partly must have other bases. In the ecological and environmental area, we must at least decide how much to compromise the immediate use of resources in order to protect ourselves. To what extent do we try to protect future generations? And to what extent do parts of our universe, especially other species, have intrinsic values themselves which are comparable to those of humans and must be protected?

Perhaps it is the human genome project and the possibility of purposeful genetic modification of humans that most cogently points out the tremendously important questions that growing knowledge forces us to face. As with other types of science, continued growth of our knowledge of genetic processes and hence the steadily increasing practicality of genetic modification probably cannot be stopped. Much of its use can be very beneficial to humanity. We can helpfully modify agricultural products. We can eliminate almost permanently certain types of diseases or even protect ourselves genetically against certain increasing environmental threats. While our present understanding is still quite limited, we must expect to be able to also modify individual human characteristics in a multitude of ways. We have in fact for some time been modifying nature's controlled evolution by medical practices which save persons and allow them to reproduce when otherwise they would not be able to do so. By medical knowledge we have increased segments of the world population which otherwise would not have been able to increase. But with the development of the knowledge of genetics these relatively minor changes and the random slow development of nature can be completely overshadowed by overt and planned genetic modification. Are we to plan what changes in humans will take place and carry them out over relatively short or over longer periods of time? Who is to determine what ways are allowed or justifiable? Does society have a right to dictate genetic change? Will the best forms of genetic modification be primarily available to individuals of wealth and the ability to buy good genetic treatments for their offspring? What represents a defect and what defects make particular genetic characteristics disposable? How far do we dare to threaten nature's more random evolutionary processes in the future development of humans? And what diversity of humans do we really want? Might we even consider developing different types of humans each for different types of jobs as ant colonies have so long ago already done—one type the super manual worker, another the super intellectual, and another the super computer operator? In the insect world, such a plan of specialization has worked well. Will some of us be tempted to think in those terms?

Biotechnology and our ability to manipulate complex molecules represent the most recent and cogent reasons for us to clearly recognize a change which has taken place in relations between humanity and the universe around us. The relatively slow modifications in our surroundings and in our species which have been dictated for us by the external universe are now superseded by the much more rapid modifications that humans are making—on their surroundings and on humanity itself. This change has developed over historic time, but has recently become acute and presents a challenge we must inevitably face—from now on, ready or not, to an uncomfortably large degree we ourselves are in charge.

The present state is the result of a very long but steadily accelerating history. The first 10 billion years after the big bang saw the cosmos prepare and produce our sun and earth. During the next $4\frac{1}{2}$ billion years life was generated on earth and developed into the diversity we now enjoy. Only about 60 million years ago the dinosaurs died off, perhaps as a result of cometary bombardment, and made room for the age of mammals. Only two or three million years ago the human species emerged. And within the last few hundred years human knowledge has brought about this remarkable change in our circumstances. No longer is our development controlled primarily by the relatively slow changes of the universe around us and evolution by somewhat random events. The time scale in which our own knowledge can change the human condition and humans themselves in major ways is now as short as a single human lifetime.

Of course, there are also some major events quite independent of humans which occur rapidly—for example, volcanic eruptions or even supernovae. And one can imagine a natural disaster or one produced by humans themselves which could be catastrophic, getting the human species back into a much more primitive state than where we are now. If such occurs at all, I think very likely it could be compensated within a rather short time from a historical point of view, such as a hundred years. Humanity could then proceed as before.

Our lives and future will of course always exist within certain boundaries determined by physical laws, or we might say by the nature of the universe. We will continue to be both supported by the cosmos and constrained by it. But within these constraints it seems unlikely that any lack of knowledge or of versatility in modifying and adapting our surroundings will limit or define human destiny. It will primarily be humans themselves who will be determining their future and the future of our environment—not external events. To an embarrassing degree, we are in charge. And it will not be so much physical limitations, but our interests and sense of values which will be primary determinants of our future.

What in fact will be our goals? The basic biological drive simply to increase population indefinitely does not now seem a very satisfying goal to most humans; it is instead a worrisome problem. I have already assumed that some of our goals will be like those of the past, for example that there will be enough human curiosity and interest in exploration and enough coherent action on the part of some fraction of our population that knowledge and ability to modify our circumstances will continue to grow. Such goals need not be universal; intellectual discovery has often been carried out by a relatively small group, or even may be an individual affair. Success in other areas can require broader collaboration, and humanity has always had difficulty in finding goals, interests, or behaviors which are universally accepted so that we can all act coherently. What instincts and purposes will in fact be dominant? Will we search for some sense of "doing good," such as human enrichment, however that may be defined? Is it development of individual powers, or group powers? Is it power over the externals we seek—to accomplish something by making a notable mark on the world, or simply to control others? Will it be physical satisfactions like eating, sex, drugs, or perhaps the ultimate passive pleasure of finding the best electrical stimuli for pleasure centers of the brain? Will our diversity and medley of values simply result in clashes and a wide variety of both purposes and outcomes?

The universe has been steadily yielding its secrets and its powers to human skill and intellect; I feel rather confident this will continue. What is crucial but not so clear about humanity's future are our values, our real desires, and our ability to act together coherently.

Reflections on My Life as a Physicist

Any attempt to talk about one's own career and the influence of religion is a delicate assignment. It is of course highly personal, and there are profound subtleties; the combination makes any easy and objective account difficult. Nevertheless, religion has had a strong effect on my life, and I shall try to indicate how it has as I make my way through this personal story.

I suppose my most formative interest has always been in nature. As a boy, I collected everything in the fields and woods—leaves, insects, bird sightings, rocks—and I looked at the sky. The wonder of the natural world inspired me and I believe much of my family. We were very fond of the outdoors. Along with wonder there was the fascination of trying to understand and to fit things together. Moreover, this interest in nature contrasted with my own lack of interest in man-made things. I was never interested, for example, in the different makes of cars. Some of my friends seemed to know every year model of every car. That seemed to me so temporary and uninteresting. Nature is such a permanent aspect of our universe, and so obviously God-made. Human artifacts, on the other hand, seemed to me trivial and temporary by comparison. Of course this was my own instinctive point of view—it can be argued that human artifacts have had their own profound effects. But I had a strong drive to make sense of the universe, and to enjoy the wonder of it. To me science is basically an attempt to understand this universe, and basically theology has the same goal.

As I look back over my own career after somewhat more than the usual three score and ten years, I'm impressed with how many times my best successes came out of failure. We probably all recognize the uncertainties in our own lives. We may think we have a chosen direction—what we want to do, and where and how we want to do it, and such a sense of direction seems to me important and satisfying. Nevertheless, we are faced with great uncertainties and the future is not well predictable.

I went to college in my native state of South Carolina, at a small Baptist university. My teachers were people of very fine quality, but they had essentially no research experience. Nevertheless, I knew by my sophomore year, when I took my first course in physics, that research in physics would be my own principal interest. I liked mathematics, but physics seemed much closer to the real world around us. I'd been interested in biology, other sciences, and languages; nevertheless physics was clearly what I wanted. And this college didn't have very much physics. So, I completed a major in physics by studying basic texts on my own and working out the problems, which were accepted as completed courses. This was perhaps some accomplishment, but I clearly still didn't know very much physics at the end of that. I applied to graduate schools. At that time, almost anyone who had reasonably decent grades could get into a graduate school, even the best ones. That wasn't so hard. The problem in the 1930s was not being admitted into a graduate school, but paying for it. I had to find a fellowship or teaching assistantship to make it through. Along with turn-downs from most of the prominent schools in physics, I was given a teaching assistantship at Duke.

Duke's a good university. However, at that time it was rather far from being a first-rank one in physics. There were some professors doing notable research, so obviously I could learn. After a year as a graduate student there, I applied for the one fellowship they had in physics. It was given to somebody else. So I decided to just finish a master's degree, take the $500 I'd saved up, and go to the very best place I knew. The sum of $500 was of course much more then than it is now, and I thought that with it I might last a year. So I set out for Caltech. Caltech is well-known as an outstanding place for physics; it was even more outstanding then. Millikan was there, Oppenheimer, Pauling, and a number of other excellent people. It was an extraordinary place. So this particular failure (to get fellowship support at Duke) took me to Caltech. I just sort of bulldozed my way there, I guess, and fortunately made it at Caltech. By midyear I received a teaching assistantship and I was all set. I could finish graduate work. Of course, my fellow-students at Caltech and professors were extremely stimulating, and that experience has been a very important one in my career.

After finishing at Caltech, I was offered a job at the Bell Telephone Laboratories. My goal was to teach and do research at a university, so when I was offered a job at Bell Telephone Laboratories I was about to turn it down. But my professors jumped on me and said, "Look, it's a job!" Jobs were very scarce, and it was the only offer I had at the time. Many of my friends who had gotten their Ph.D.s were teaching in high schools or took jobs in the oil fields to work in seismology because at that time the oil fields represented one place where there were jobs. So, after some days of uncertainty about how long I might wait to find a good alternative, I took the advice of my professors and went to the Bell Telephone Laboratories. I knew there were good people there, and I knew some of the outstanding work that they had done. But it was not exactly what I had in mind as a way to explore fundamental physics. Again, in spite of my initial orientation toward a university and my failure to find such a job, the Bell Telephone Laboratories turned out to be extremely important and helpful to me, and later to my career in academic science.

Being part of a commercial lab rather than a university was not the only deflection of my plans. For about fifteen months after I arrived at Bell Labs, I did physics research as I had been hired to do. But one Friday I was called in by the head of the laboratory and told, "On Monday you start working on the design of radar bombing systems. The war is coming on, and that's your assignment!" So over the weekend I converted suddenly from physics to engineering. On the one hand, I found the arbitrariness of this reassignment very objectionable. On the other hand, I expected that soon almost everyone would have to serve the war effort, and concluded that perhaps this must be my part.

During the war years we worked on some of the early analog computers, navigation and bombing systems, and of course radar. Now, that turned out to be exceedingly important to me; much of my subsequent work has grown out of it. How? First, I learned electronics. Most physicists didn't know electronics very well at that time. Also I learned about microwaves, and the early stages of computing. Perhaps most important of all, in the last years of the war we were learning how to produce and work with shorter and shorter wavelengths in the microwave region, in order to have better and better directivity of the radars and from this I glimpsed my future research.

In spite of early success with some radar navigation and bombing systems, we were told, "Okay, you've made a system with microwaves of 3 centimeters wavelength. Now the decision has been made to push for a system with wavelengths down to one and a quarter centimeters." Naturally I was annoyed that, having worked intensively to build and demonstrate one elaborate system, it was scrapped in favor of a next one. While planning the next system, I read an informal manuscript of Professor Van Vleck that pointed out, somewhat loosely because not much was known, that water vapor absorbed radiation at this shorter wavelength range. I looked at the problem as well as I could, and it seemed clear to me that water vapor was likely to absorb waves at $1\frac{1}{4}$ centimeters wavelength so much that they wouldn't go far in wet air. The course of the war was changing, with the Japanese conflict becoming its most important aspect. The Pacific was very wet, and I concluded these microwaves wouldn't go very far. After careful preparation, I presented this to various authorities above me. But I was young; they wouldn't listen and decided to go ahead.

The radar was built, and it didn't work for the reason I had predicted. This particular failure got me thinking about the interaction between microwaves and molecules, a field that was practically nonexistent. I was able to recognize that there was a way of determining molecular properties very precisely, and a spectacular kind of spectroscopy. The accuracy could be beyond that ever before attained, with very narrow spectral lines, very high accuracy, and precise determination of many types of details about molecules. I wrote a memorandum trying to persuade Bell laboratories to let me do this immediately after the war. Fortunately, after a short delay they agreed, and that started me on microwave spectroscopy. The failed radar turned out to be very useful because scrapped parts of the $1\frac{1}{4}$ centimeter radar were exceedingly cheap and useful in this research! From it we found out a great deal about properties of nuclei and of molecules, as well as new physical ideas.

Along with microwave spectroscopy, I had become interested in radio astronomy. Radio astronomy had been discovered accidentally at Bell Labs by an engineer who, of all things, found a principle source of radio noise to originate in the center of our galaxy! Nobody understood this phenomenon very well then. Based on my fascination with astronomy and these developments, I went to Palomar Observatory to discuss my potential involvement in radio wave astronomy with a highly respected former professor whom I knew from Caltech and who was head of the Observatory. Bell Labs seemed like an ideal place where astronomy might be studied with radio or microwaves. But he said, "Sorry to disappoint you, but I don't think radio waves are ever going to tell us anything about astronomy." That was in 1945 or 1946, and of course only about five years later radio astronomy had become a rapidly developing world-wide enterprise. It is now an exceedingly important aspect of astronomy. Nevertheless, his remarks were enough to turn my attention back to microwave spectroscopy, which seemed to me an equally promising field. My work in this field at Bell Laboratories went well, but officials there weren't all that interested in it because it was not clear to them how it would contribute to communications. I tried to persuade them to the contrary, and they were generally supportive but not committed enough to expand the field.

Fortunately my work progressed well enough that in 1947 I was offered a suitable professorship at Columbia University, and I accepted. Microwave research grew steadily during these years. We were working at that time with waves about one centimeter long. The shorter the waves the more strongly the interaction with molecules. Hence I was highly interested in working with still shorter waves, and doing my best to discover ways of developing this field. Eventually I was asked to chair a national committee to determine how to distribute funds designated by the Navy for research on short microwaves. The Navy was interested in developing the field primarily for exploratory reasons, but it was also quite supportive of basic science at that time. Our committee listened to many ideas, came up with some of its own, and wrote a few papers to try to stimulate developments. But after about a year and a half of trying, nothing very fruitful had turned up.

We were about to have an important meeting in Washington in order to review our work. I awoke early on the day of the meeting, perhaps out of my concern and frustration, and went out to nearby Franklin Park. It was a beautiful morning and the azaleas were out. I sat on a bench and mused over why we had failed. Why had we failed to find a way to produce waves shorter than a few millimeters? One reason clearly was the limit in our capacity to make things small enough with man-made instruments. So far as we understood, shorter waves required smaller devices. And furthermore, such small devices could not dissipate the power required to produce the short waves. I concluded that if we were to get very far, we had to do it somehow with molecules—molecules already made by nature with sharp resonant frequencies and other wonderful properties. But the second law of thermodynamics says it can't be done that way. Suddenly I realized we didn't have to obey the second law of thermodynamics. That's a law applying only when molecules have their usual random behavior which can be described by temperature equilibrium. It's a powerful law, but in fact we didn't have to have equilibrium! That was a sudden recognition on my part! It was not a new idea—most scientists knew the second law of thermodynamics doesn't always apply, yet we almost always applied it. And I saw immediately how to put molecules in an unusual state where they weren't obeying the second law of thermodynamics, so they could amplify and produce short wave radiation. I would also use a resonator, another device quite familiar to me from my Bell Laboratories work. In principle, it would not yield a very large amount of power. But I decided to try it, with the important help of a young postdoc and graduate student.

We worked together for a couple of years. Then, one day the former chair of the department, Professor Rabi, and the present chair, Professor Kusch, both very well-known physicists, came into my office. They sort of banged on the table and said, "Look, you know that is not going to work. We know it's not going to work. You're wasting money and time. You ought to stop it!" Among scientists, this is highly unusual.

Well, I simply told them, "No, I still think there is a reasonable chance that it will work." (How thankful I was that I had tenure!) And fortunately, three months later we had it working. Professor Kusch generously congratulated me later, saying he should know that I probably understood what I was doing better than he. Microwave energy was being generated from molecules with the new device we called a maser, for *M*icrowave *A*mplification by *S*timulated *E*mission of *R*adiation. A significant breakthrough had occurred, though new refinements and developments were to follow.

During that same period, I vividly recall an occasion when another faculty member, an atheist but good physicist friend who knew of my religious faith, facetiously asked me, "Did God ever really help you in the laboratory?" And I said, "Yes, I think so." Well, that stopped him. He didn't ask the second question, "How?" I think he was too flabbergasted by my reply to continue. I'll try to indicate at least some ways in which I think my reply was true, even though one may still wonder.

At about that time I went on sabbatical, once again to consider new directions. The idea of entering astronomy resurfaced, but various things drew my attention back to development of the maser, and shortly after returning from sabbatical I decided to face the problem of overcoming the barriers to still shorter wavelengths. We had made microwaves, but I wanted to produce still shorter waves, even down into the infrared region, my original goal. They would be still more valuable tools for research. Instead of the Washington experience where on the spur of the moment I contemplated problems in the beauty of a park, this time I simply decided to sit at my desk and plough through the problem. I played with the equations until suddenly I realized from them that it would be just as easy to go right on to still shorter waves—light waves. An advantage was in part that we had all kinds of apparatus to deal with light waves, but the far infrared region just shorter than microwaves was not yet well developed. Bell Telephone Laboratories in the meantime had asked me to come back on a consultant basis, and while I was there I talked with Arthur Schawlow about these ideas. We had already worked together extensively and also had family ties. Schawlow expressed great interest in the project of obtaining a very short-wave source from the maser, and in fact had already been wondering about the same problem himself. So the stage was set for collaborative work.

Schawlow contributed a very important idea about how to produce a convenient resonator where the waves could be trapped for some time to allow them to interact strongly with the molecules or atoms which were to increase their intensity. Quite simply, his idea involved two mirrors so that the light waves could go back and forth between them. While there were sophisticated ramifications we worked out, it was a very simple idea; once an idea becomes recognized, it is likely to seem simple in retrospect. Beforehand, it's unknown. The concept is this: light waves go back and forth through these molecules or atoms, and each atom is stimulated by it to give a little energy to the light wave. The light is also in essentially perfect phase. That is, it maintains the same wavefront as it travels back and forth. It can go back and forth a number of times, stimulating molecules or atoms to release their energy, and creating a strong beam. If it goes back and forth indefinitely, the beam increases until the rate it loses energy by, say, leaking through the mirrors or by imperfect reflections is just equal to the rate it gains energy from the molecules. There is then a steady oscillation, and a highly directed beam is emitted. This was the laser, *L*ight *A*mplification by *S*timulated *E*mission of *R*adiation.

One of the greatest aspects of science is that once we understand something, it can be explained to others and soon it's a property of the human race. And for those who understand, it often seems rather simple, In fact, one wonders why these ideas didn't come together 25 years earlier; none of the principles were unknown even that early. And why did it happen just then? Of course, I had a pertinent general orientation and direction: I wanted to be at a university, to do research, I was interested in physics, and I had a reason to try to produce short waves. But why this all happened as it did still seems a happy mystery, even as I look back on the explainable individual steps.

The general idea of the maser and laser can indeed now seem rather simple; however, there were matters of principle which people continue to argue with me about. For instance, before the maser was operating some very important physicists refused to believe that it would give very pure frequencies, as I thought I could prove. One of them bet me a bottle of scotch. Another didn't believe our ideas, argued strongly against them, and after they worked he seemed to avoid me. But again, it wasn't really a matter of *new* principles as much as the way they were understood and put together.

At this point, I was becoming fairly well known, and for some reason—chance, I think—I was asked to go to Washington. It was the era of the missile gap, and Sputnik, with the U.S. late in being able to put anything into orbit. There weren't many scientists in Washington; most scien-

tists wouldn't want to go (at least one or two people had been asked before I was). The proposed position for me was Vice President and Director of Research for the Institute for Defense Analysis. The Institute was a
non-profit "think-tank" with a very important role, run by five or six
prominent universities on the East Coast, Columbia University being one
of them. It managed what was known as the Weapons Systems Evaluation
Group. We had to pick the right people who would be responsible for ana-
lyzing how and whether a weapon worked and its effectiveness. We also
advised a new organization, the Advanced Research Projects Agency,
whose aim was to consider what could be done in space, and to help initiate new ideas and technologies of importance to national security. We also
advised the State Department on arms control problems. Accepting this
position was one of the most difficult decisions for me, for it shifted my
focus from very interesting research, including at that time trying to make
the first laser operate, to occupation with political and administrative con-
cerns. But I felt it was a duty. Somebody had to do it, and there weren't
many well-known scientists who would. The government desperately needed
help, I felt, to make sensible decisions in matters involving advanced technical applications. So with my wife's consent, we moved to Washington.

As an illustration of the dilemma the U.S. was in, only a few days after
I arrived I met with Allen Dulles, then head of the CIA (and brother of
John Foster Dulles) and he tried to give me the story of what was known
about Soviet intercontinental missiles. It was a lengthy presentation of all
the highly classified evidence we had. Finally he asked me, "What do you
think? How many might the Soviet Union have?" There was great concern that the Soviets might have tens, or maybe hundreds of intercontinental missiles and we had none. I had to answer, "I don't think we know.
They might have zero, they might have fifty. I see no clear evidence."

Since then, there have been major advances, as you probably know.
With satellites we can now tell a great deal more. Satellite surveillance,
which we scrambled to achieve, has helped a great deal to provide important information and stabilize the situation. Suspicion about what an op-
ponent may be up to without any real knowledge can very well lead to
over-reaction, and destabilization with frightening scenarios.

After two years this part of my career was coming to an end and I was
eager to get back to research. But M.I.T. came along and wanted me to be
provost. They made it clear that they needed a new president in about four
or five years. While I do not object to administration, I'm not all that interested in it either. Still, I felt M.I.T. was one place where it would make
sense for a scientist to do administration. So I went there as provost. But,
ironically, when the then president retired, they picked somebody else, not

a scientist or engineer. This was a personal disappointment, but on the other hand, it allowed me to get back to science. I had other offers to head academic institutions, but given my reluctance to do administration, I felt that only concentrating on science and engineering made sense for me. Along with my years in Washington, the M.I.T. experience had given me experience in administration and public affairs, but now it was time for another transition and to take advantage of another, perhaps fortunate, failure.

I came to California, planning to move toward astronomy and choosing California as the right place for it. One of the first things I wanted to do was to look for molecules in interstellar space. I'd written a paper about five years earlier based on my work in microwave spectroscopy and asserting there was a reasonable chance that certain molecules existed in interstellar space and might be detected. Several astronomers thanked me, showed great interest, and said that searches should be made. But, strangely enough, no one was convinced enough to do it. Most astronomers persuaded themselves that the molecules wouldn't really be there, so nobody looked. For example, when I came to Berkeley and started to prepare for a search for molecules, the head of the astronomy department told me those molecules simply can't be there. It was thought there are no clouds dense enough to form the molecules and they would in any case be destroyed by ultraviolet radiation, so they just couldn't be there. I felt there was a case to the contrary, and fortunately did have some helpful colleagues in the astronomy department. So after some months of preparation we looked and, sure enough, we found ammonia, and then water, which were in fact common in certain parts of interstellar space. About a hundred different molecules have now been found, including all those thought be to be the most important in initiating life, floating in interstellar space and waiting to condense into stars and planets. Again, this work was in some sense an outgrowth of the fact that I had to develop radar bombing systems. What one does of course grows out of past experience. And of all things, this new astronomy had its roots in my somewhat arbitrary assignment during the war.

The field of molecular astronomy is now developed into a very important one and is very active. So I have moved to other things. My taste is not for staying in fields that are very popular. I prefer to do something that is a little different, to explore paths which seem promising to me but others have overlooked. Most recently I've been doing infrared astronomy, now in particular infrared interferometry. It is very complex and difficult, but I am hopeful that it too will give birth to another interesting and active field of research and will teach us much about the stars.

Now that I have given you this brief story, I shall try to consider how the things that formed me as a person contributed to the course I have taken. Certainly much of it is family background—attitudes toward life that came about from my family and others. Let me illustrate one of the family characteristics. While in my role as professor in Berkeley I'm generally regarded as a political conservative. Actually, I think of myself as a moderate who examines carefully the problems with what others are saying. So yes, I have objected to some of the political trends in Berkeley, and made that plain. But my friends in government and industry regard me as somewhat tainted by Berkeley radicalism, because I do the same thing with them: I accept where I think they're right and try to point out where I think they're wrong. As I look back at my family history, perhaps I can spot this trait in my ancestor, William Bradford, a Mayflower pilgrim who came to America because of his objections to the then contemporary society. Another ancestor, William Screven, was a Baptist who was run out of Maine by Congregationalists because his religious views did not coincide with their norm. Being determined and a man of conviction, he went South and started the first Baptist church there. Still another family member, a newspaper editor in Charleston, South Carolina, was an abolitionist before the Civil War. When the tension built up toward the Civil War he lost his position because of it. My parents, too, were people who stepped aside from popular movements and bandwagons in order to represent what seemed to them right, or true.

I regard such instincts as deeply rooted in religious background and a critical aspect of both creative science and real religion. It reflects also our respect for individualism—a certain resistance toward easy assent to a collective will, or the popular thing. Worldly things and popular ideas are not to be disdained, but we must be able to see and understand in different ways. In addition to willingness to take positions which aren't necessarily popular, there is another thing I believe a religious outlook provides, and which is an important aspect of science. One needs to know how to disagree with people, but also to do it with humility. It is important to be right, but to achieve that we must try to weigh the truth with humility. Those who disagree with us are worth listening to; they are a key ingredient to self-examination and recognition of our own errors. Thus the ability to stand apart from the crowd as one looks at life, combined with personal humility in the pursuit of reality and objectivity, are two very important scientific qualities and, I believe, also inherent parts of enlightened religion.

Let me comment on another aspect of a scientist's life, namely, public work. Clearly, somebody has to contribute public service and participate in government. I've always felt that the government sector needs scientists. In my case, I don't especially enjoy it, but I can tolerate it and perhaps it is partly my religious orientation that makes me believe it's part of my duty to society to serve in this way. Connections with government have not always been popular here in Berkeley, and were strongly unpopular during the Vietnam war and for some time thereafter.

In 1981 I was asked by the Secretary of Defense to advise the administration on what to do about the MX missile, in particular its deployment. This was shortly after the Reagan administration had come into office, and hence a time to look at things freshly. I've always found a transition in government to be the best time to shake things up and do something different. Once a government is in power for awhile, it usually doesn't really want to change. The Secretary of Defense agreed that I should select members of an advisory committee, though he was to retain veto power of any particular person he felt would be too objectionable because of publicly taken political positions. He agreed on still other conditions of importance to me, and assured me that our considerations would be free of political pressure. We were to make judgments about the MX missile in terms of safety and usefulness to world stability and peace.

Just before I took on this task the minister of my church preached a sermon against the MX missile. He knew what I was about to do, so I asked him whether he was trying to tell me something. "Oh, no, I wasn't really talking about you," he said. I replied, "Well, you perhaps have an opinion. What is it?" He answered, "I don't think you ought to do this. The MX missile is terrible, you shouldn't be involved with it." My reply was a question. "Suppose there was somebody that you even thought was a criminal, who came to you saying, 'Look, I'm in trouble, I need advice and wonder if you would talk with me and help.' Would you refuse?" "No," he said. So I responded, "The U.S. government says they want my help and advice in what seems to involve critical decisions. Should I say no?" He concluded abruptly by saying in effect, "Well, just go your way and do whatever you think; but I wouldn't have anything to do with the MX missile." Of course, I did it.

That was a fairly difficult, tense time. I chose a committee representing a wide variety of viewpoints and backgrounds. It included the now well-known Brent Scowcroft and other high-ranking military figures of his caliber as well as academic scientists and analysts. The MX missile system was scheduled by the Carter administration to be 200 missiles, each having 10 warheads. They would be deployed in 2,000 holes scattered throughout Utah and Nevada; 2,000 silos and 200 missiles each with 10 warheads. Our group studied how useful, effective, and safe the system would be, and what were the alternatives, then made its report. There was a majority opinion and a minority opinion and of course I was on the minority side, as usual. There were tough arguments on all points, and frequently we would be in discussions all day. Things would seem to be going the wrong way. Frankly, at night I would pray, and the next morning, things seemed to be better. For some reason the Committee's thoughts seemed to clear up in the right direction the next morning. The entire committee finally agreed against recommending construction of the 2,000 silos. The majority conclusion was to build the 200 missiles, but to deploy them more simply rather than all over Utah and Nevada. The recommendation of the minority group, two academic types and a navy admiral, was against this. They recommended maintaining a modest production capability but building only those already under production. This might persuade the Russians of our capacity and determination and discourage them from further deployment of large multiple warhead missiles, which they already had in some numbers. The minority also recommended against deployment. As it turned out, I argued the case before the President and his cabinet and Reagan's decision was to build only 40 missiles and 40 single silos. A remarkable turnaround. One could argue whether that was right or wrong. But it was a remarkable turnaround, and I felt that good things were at work. I still believe the MX missile's appropriate function was to be in reserve as a deterrent to further Soviet deployment, and it served this purpose.

I've given you a sense of the course of my career—some key decisions, directions, and perhaps happy failures. You may well ask, "Just where does God come into this?" Perhaps my account may give some answer but to me that's almost a pointless question. If you believe in God at all, there is no particular "where;" He's always here—everywhere. He's in all of these things. To me God is personal, yet omnipresent—a great source of strength, Who has made an enormous difference to me. When my atheistic friend asks, "What has he done for you?" What can I say? I look at what's happened to me, and think that all of those things are what He's done. Acceptance of His role is perhaps a bit analogous to our feelings about some form of free will. Free will is not logical or possible in terms of our present understanding of science. Yet we have an almost irrepressible sense that we act with free will and would probably be quite different without this sense. We can hope our presently limited understanding will improve and some day clarify these instinctive feelings. And so while one can wonder whether such a figure as God can in fact exist, we may sense him strongly—both at the moment and in reflecting on the events of a lifetime.

Acknowledgments

Material published here is based on the following sources. The author is very grateful for permission from these to reprint previously published material.

QUANTUM ELECTRONICS AND SURPRISE IN DEVELOPMENT OF TECHNOLOGY is a talk given before the American Association for the Advancement of Science and published in *Science*.

ORIGINS OF THE MASER AND THE LASER is based on a talk given on the occasion of the Fortieth Anniversary of the Joint Services Electronics Program in 1987, which was published in its *Proceedings*.

ESCAPING STUMBLING BLOCKS IN QUANTUM ELECTRONICS comes from a talk at the 1984 centennial celebration of the Institute of Electrical and Electronics Engineers published in the *IEEE Journal of Quantum Electronics*.

MICROWAVE SPECTROSCOPY is from my Sigma Xi national lectureship in 1951. The lecture was published in Sigma Xi's journal, the *American Scientist*.

MASERS is a lecture commemorating the tenth anniversary of M.I.T.'s Lincoln Laboratories, published in a book of such lectures which was edited by Lincoln Laboratories' director, Carl Overhage.

MESSAGES FROM MOLECULES IN INTERSTELLAR SPACE is based on a lecture given on the occasion of the award of the Niels Bohr International Gold Medal and published by Dansk Ingeniørforening.

SCIENCE, TECHNOLOGY, AND INVENTION: THEIR PROGRESS AND INTERACTIONS is from a talk at the National Academy of Sciences' Symposium of that name. It was published in *Science*.

THE POSSIBILITIES OF EXPANDING TECHNOLOGY was part of a discussion honoring the centennial of the Institute of Electronics and Electrical Engineers which was published in *Electrical Engineering: The Second Century Begins* by the IEEE Press.

THE ROLE OF SCIENCE IN MODERN EDUCATION is based on a lecture given at the dedication of the Hugh Miller Willet Science Center of Mercer University, published in the *Southern Baptist Educator*.

RESEARCH LABS: VARIETY AND COMPETITION results from a Conference on Research organized by the American Academy of Arts and Sciences at its Paris center. It was published in *Daedalus*.

HOW AND WHY DID IT ALL BEGIN? comes from an article in the *Journal of the American Scientific Affiliation*.

THE CONVERGENCE OF SCIENCE AND RELIGION originated as a talk to a men's group at Riverside Church in New York City. First published in IBM's magazine *Think*, the present version comes from M.I.T.'s *Technology Review*.

WHAT SCIENCE SUGGESTS ABOUT US is based on a talk at a Biannual Meeting of the Association of Theological Schools, and was published in *Theological Education*.

WHY ARE WE HERE? WHERE ARE WE GOING? is adapted from a lecture at a meeting of L'Institut de la Vie World Center for Human Destiny, and is to be published by the American Institute of Physics in a volume honoring Victor Weisskopf.

REFLECTIONS ON MY LIFE AS A PHYSICIST is a lecture given at the Center for Theology and Natural Sciences and published in the *CTNS Bulletin*.

Index

About the Author

A scientist whose inventions have permanently altered the landscape of modern technology and daily life, Charles H. Townes won the 1964 Nobel Prize in physics for two major breakthroughs. He conceived both the laser and the maser; he has the original patent on the maser, and with Arthur Schawlow also on the laser. He has also made major contributions to molecular physics and the advancement of astrophysics, among them his discovery of stable molecules in galactic clouds.

Born in Greenville, South Carolina, Dr. Townes graduated from Furman University, earned his master's degree in physics from Duke University, and his Ph.D. at the California Institute of Technology. He worked initially at the Bell Telephone Laboratories, then taught at Columbia University, spent some time in Washington, returned to academic work at the Massachusetts Institute of Technology, and later became University Professor at the University of California. During the latter part of his career he has been an active governmental advisor, including service as chairman of the technical advisory committee for the Apollo Program.

Among his many honors, Dr. Townes received the National Medal of Science in 1982. He is also a recipient of the Niels Bohr International Gold Medal and is a member of the National Academy of Sciences, the Royal Society of London, the National Inventors Hall of Fame, and the Engineering and Science Hall of Fame.

More recently Dr. Townes has developed a pair of moveable telescopes for obtaining very high angular resolution of astronomical objects. He is presently University Professor Emeritus at the University of California.